T0186987

WJEC
Biology
for A2 Level
Revision Workbook

Neil Roberts

Published in 2022 by Illuminate Publishing Limited, an imprint of Hodder Education, an Hachette UK Company, Carmelite House, 50 Victoria Embankment, London EC4Y 0DZ

Orders: please contact Hachette UK Distribution, Hely Hutchinson Centre, Milton Road, Didcot, Oxfordshire, OX11 7HH. Telephone: +44 (0)1235 827827. Email: education@hachette.co.uk. Lines are open from 9 a.m. to 5 p.m., Monday to Friday. You can also order through our website: www.hoddereducation.co.uk

British Library Cataloguing in Publication Data

A catalogue record for this book is available from the British Library

ISBN 978-1-912820-40-5

Printed by Ashford Colour Press, UK

Impression 3

Year 2024

Hachette UK's policy is to use papers that are natural, renewable and recyclable products and made from wood grown in well-managed forests and other controlled sources. The logging and manufacturing processes are expected to conform to the environment regulations of the country of origin.

Every effort has been made to contact copyright holders of material produced in this book. Great care has been taken by the authors and publisher to ensure that either formal permission has been granted for the use of copyright material reproduced, or that copyright material has been used under the provision of fairdealing guidelines in the UK – specifically that it has been used sparingly, solely for the purpose of criticism and review, and has been properly acknowledged. If notified, the publisher will be pleased to rectify any errors or omissions at the earliest opportunity.

Editor: Geoff Tuttle
Design: Nigel Harriss
Layout: John Dickinson

Cover image: Shutterstock.com/Richard A McMillin

Acknowledgements

For Isla and Lucie.

Thank you Louise for your patience and support which made this book possible.

The author would also like to thank the editorial team at Illuminate Publishing for their support and guidance.

Picture Credits

p56 (r) (C) Hugo Ahlenius, https://www.grida.no/resources/8324; p56 (ml) Adapted from http://www.projectarkfoundation.com/animal/bornean_orangutan; p56 (bl) Adapted from http://www.projectarkfoundation.com/animal/bornean_orangutan; p66 Shutterstock/Anna Jurkovska; p67 Shutterstock / Anna Jurkovska; p78 National Center for Biotechnology Information / Public domain; p79 National Center for Biotechnology Information / Public domain; p120 (t) Aldona Griskeviciene / Shutterstock; p120 (bl) Shutterstock / skys.co.jp; p120 (br) Shutterstock / kio88; p121 Shutterstock / DragoNika; p138 © Courtesy of Innovative Care, 2021; p155 DR. GEORGE CHAPMAN, VISUALS UNLIMITED / SCIENCE PHOTO LIBRARY; p172 (bl) Shutterstock / Pete Niesen; p172 (br) Shutterstock / Pavaphon Supanantananont; p172 (t) Shutterstock / mlopez; p179 Shutterstock / BioMedical; p182 Copyright © 2015 Osteoarthritis Research Society International. Published by Elsevier Ltd. All rights reserved; p67 © Karen Hart

Other illustrations © Illuminate Publishing

Contents

How to use this book

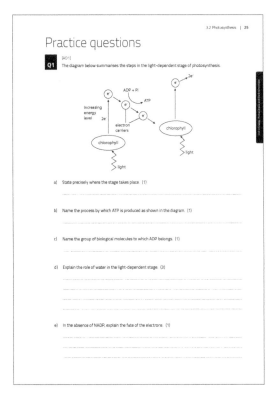

Each topic section of the guide begins with around four to eight examination-style practice questions focussed on each topic. The assessment objective being tested is identified for you. The real exam questions will frequently draw from several topics, and from AS/Year 1, so it is important, once you feel confident with the Practice questions, you move on to complete past papers (one is provided for each section in this book), and many are available from the exam board website.

This is then followed by some examples of actual student answers to questions. In each case there are two answers given; one from a student (Lucie) who achieved a high grade and one from a student who achieved a lower grade (Ceri). We suggest that you compare the answers of the two candidates carefully; make sure you understand why one answer is better than the other. In this way you will improve your approach to answering questions.

Examination scripts are graded on the performance of the candidate across the whole paper and not on individual questions; examiners see many examples of good answers in otherwise low-scoring scripts. The moral of this is that good examination technique can boost the grades of candidates at all levels.

This exam preparation guide has been designed to work alongside the Study and Revision Guide, also available by the same author and published by Illuminate Publishing.

Assessment objectives

Examination questions are written to reflect the assessment objectives (AOs) as laid out in the specification.

The three main skills that you must develop are:

AO1: Demonstrate knowledge and understanding of scientific ideas, processes, techniques and procedures.

AO2: Apply knowledge and understanding of scientific ideas, processes, techniques and procedures.

AO3: Analyse, interpret and evaluate scientific information, ideas and evidence, including in relation to issues.

In both written examinations you will also be assessed on your:
- Mathematical skills (minimum of 10%)
- Practical skills (minimum of 15%)
- Ability to select, organise and communicate information and ideas coherently using appropriate scientific conventions and vocabulary.

These are indicated against each topic question together with mathematical skills (M) and practical skills (P) where they occur.

In any one question, you are likely to be assessed on all skills to some degree. It is important to remember that only about a third of the marks are for direct recall of facts. You will need to apply your knowledge, too. If this is something you find hard, practise as many past paper questions as you can. Many examples come up in slightly different forms from one year to another.

Your practical skills will be developed during class-time sessions and will be assessed in the examination papers. These could include:
- Plotting graphs
- Identifying controlled variables and suggesting appropriate control experiments
- Analysing data and drawing conclusions
- Evaluating methods and procedures and suggesting improvements.

Understanding AO1: Demonstrate knowledge and understanding

You will need to demonstrate knowledge and understanding of scientific ideas, processes, techniques and procedures. Approximately 27% of the available marks set on the A2 exam papers are for recall of knowledge and understanding.

Common command words used here are: state, name, describe, explain.

This involves recall of ideas, processes, techniques and procedures detailed in the specification. This is content you should know.

A good answer is one that uses detailed biological terminology accurately and has both clarity and coherence.

If you were asked to describe and explain how electrophoresis produced the results seen in a gel, you might write:

'DNA moves towards the positive electrode through the gel. Smaller fragments move further.'

This is a basic answer.

A good answer needs to be more detailed. For example,

'DNA is attracted to the positive electrode due to the negative charge on its phosphate groups. Smaller fragments find it easier to migrate through the pores in the gel and so travel further than larger fragments during the same time. The size of fragment can be estimated by running a DNA ladder which contains fragments of known size alongside the sample.'

Understanding AO2: Applying knowledge and understanding

You will need to apply knowledge and understanding of scientific ideas, processes, techniques and procedures:

- In a theoretical context
- In a practical context
- When handling qualitative data (this is data with no numerical value, e.g. a colour change)
- When handling quantitative data (this is data with a numerical value, e.g. mass/g).

45% of the available marks on the A2 exam papers are for application of knowledge and understanding.

Common command words used here are: describe (if it's unfamiliar data or diagrams), explain and suggest.

AO2 tests applying ideas, processes, techniques and procedures detailed in the specification to unfamiliar situations including using mathematical calculations and interpreting the results of statistical tests.

If you were asked to describe the effects of a weed-killer on non-cyclic photophosphorylation explaining why cyclic photophosphorylation was unaffected given the information that the weed-killer blocks electron flow from Photosystem II to the electron carrier, you might write:

'It stops electrons moving out of Photosystem II into the electron carrier so electrons can't pass to Photosystem I.'

This is an incomplete answer and does not explain why cyclic photophosphorylation is unaffected.

A good answer would say:

'The weed-killer stops electrons from Photosystem II being moved to Photosystem I, which prevents the reduction of NADP to reduced NADP. Photolysis of water cannot occur. Cyclic photophosphorylation is not stopped because the electrons are still able to pass from Photosystem I and return back to Photosystem I.'

Describing data

It is important to describe accurately what you see, and to quote data in your answer.

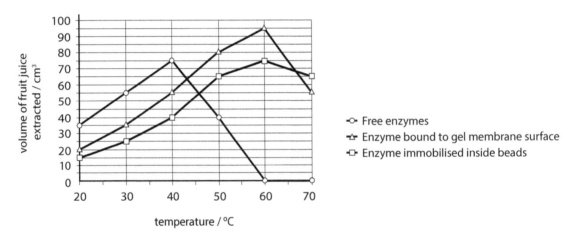

If you were asked to compare the volume of juice produced when using enzymes bound to the gel membrane surface compared to the enzyme immobilised inside the beads, you might write:

'The volume of juice extracted increases with temperature up to the optimum temperature of 60 °C in both enzymes. Above this, the volume of juice decreases.'

This is a basic answer.

A good answer needs to be both accurate and detailed. For example:

'Increasing temperature causes the volume of fruit juice extracted to increase up to 60 °C. The volume of juice collected is higher up to 60 °C with the enzyme bound to the gel membrane, peaking at 95 cm³ compared to 75 cm³ for the enzyme immobilised inside the beads. Above 60 °C the volume of fruit juice extracted decreases, but this is more noticeable for the enzymes bound to the gel membrane surface which decrease by 40 cm³ compared to just 10 cm³ for the enzyme immobilised inside the beads.'

If you were also asked to explain the results, a basic answer would include reference to *'increased kinetic energy up to 60 °C, and denaturing enzymes above 60 °C'*. A good answer is one that uses detailed biological terminology accurately and has clarity and coherence. A good answer would also include reference to *'increased enzyme–substrate complexes forming up to 60 °C'* and would include that *'above 60 °C, hydrogen bonds break, resulting in the active site changing shape so fewer enzyme–substrate complexes could form'*.

Mathematical requirements

A minimum of 10% of marks across the whole qualification will involve mathematical content. Some of the mathematical content requires the use of a calculator, which is allowed in the exam. The specification states that calculations of the mean, median, mode and range may be required, as well as percentages, fractions and ratios. The additional requirements included at A level are shown in bold.

You will be required to process and analyse data using appropriate mathematical skills. This could involve considering margins of error, accuracy and precision of data.

Concepts	Tick here when you are confident you understand this concept
Arithmetic and numerical computation	
Convert between units, e.g. mm^3 to cm^3	
Use an appropriate number of decimal places in calculations, e.g. for a mean	
Use ratios, fractions and percentages, e.g. calculate percentage yields, surface area to volume ratio	
Estimate results	
Use calculators to find and use power, exponential and logarithmic functions, e.g. estimate the number of bacteria grown over a certain length of time	
Handling data	
Use an appropriate number of significant figures	
Find arithmetic means	
Construct and interpret frequency tables and diagrams, bar charts and histograms	
Understand the principles of sampling as applied to scientific data, e.g. use Simpson's Diversity Index to calculate the biodiversity of a habitat	
Understand the terms mean, median and mode, e.g. calculate or compare the mean, median and mode of a set of data, e.g. height/mass/size of a group of organisms	
Use a scatter diagram to identify a correlation between two variables, e.g. the effect of lifestyle factors on health	
Make order of magnitude calculations, e.g. use and manipulate the magnification formula: magnification = size of image / size of real object	
Understand measures of dispersion, including standard deviation and range	
Identify uncertainties in measurements and use simple techniques to determine uncertainty when data are combined, e.g. calculate percentage error where there are uncertainties in measurement	
Algebra	
Understand and use the symbols: $=, <, \ll, \gg, >, \propto, \sim$.	
Rearrange an equation	
Substitute numerical values into algebraic equations	
Solve algebraic equations, e.g. solve equations in a biological context, e.g. cardiac output = stroke volume × heart rate	
Use a logarithmic scale in the context of microbiology, e.g. growth rate of a microorganism such as yeast	
Graphs	
Plot two variables from experimental or other data, e.g. select an appropriate format for presenting data	
Understand that $y = mx + c$ represents a linear relationship	
Determine the intercept of a graph, e.g. read off an intercept point from a graph, e.g. compensation point in plants	
Calculate rate of change from a graph showing a linear relationship, e.g. calculate a rate from a graph, e.g. rate of transpiration	
Draw and use the slope of a tangent to a curve as a measure of rate of change	
Geometry and trigonometry	
Calculate the circumferences, surface areas and volumes of regular shapes, e.g. calculate the surface area or volume of a cell	

Understanding AO3:
Analysing, interpreting and evaluating scientific information

This is the last and most difficult skill. You will need to analyse, interpret and evaluate scientific information, ideas and evidence, to:

- Make judgements and reach conclusions
- Develop and refine practical design and procedures.

Approximately 28% of the available marks on the A2 exam papers are for analysing, interpreting and evaluating scientific information.

Common command words used here are: evaluate, suggest, justify and analyse.

This could involve:

- Commenting on experimental design and evaluating scientific methods
- Evaluating results and drawing conclusions with reference to measurement, uncertainties and errors.

What is accuracy?

Accuracy relates to the apparatus used: How precise is it? What is the percentage error? For example, a 5 ml measuring cylinder is accurate to ±0.1 ml, so measuring 5 ml could yield 4.9–5.1 ml. Measuring the same volume in a 25 ml measuring cylinder which is accurate to ±1 ml would yield 4–6 ml.

Calculating % error

It's a simple equation: accuracy/starting amount × 100. For example, in the 25 ml measuring cylinder the accuracy is ±1 ml so the error is 1/25 × 100 = 4%, whereas in the 5 ml cylinder the accuracy is ±0.1 ml so the error is 0.1/5 × 100 = 2%. Therefore, for measuring 5 ml it is better to use the smaller cylinder as the % error is lower.

What is reliability?

Reliability relates to your repeats. In other words, if you repeat the experiment three times and the values obtained are very similar, then it indicates that your individual readings are reliable. You can increase reliability by ensuring that all variables that could influence the experiment are controlled, and that the method is consistent.

Describing improvements

If you were asked to describe what improvements could be made to the reliability of the results obtained from an experiment extracting apple juice, you would need to look closely at the method and apparatus used.

Q: Pectin is a structural polysaccharide found in the cell walls of plant cells and in the middle lamella between cells, where it helps to bind cells together. Pectinases are enzymes that are routinely used in industry to increase the volume and clarity of fruit juice extracted from apples. The enzyme is immobilised onto the surface of a gel membrane, which is then placed inside a column. Apple pulp is added at the top, and juice is collected at the bottom. The process is shown in the diagram. Describe what improvements could be made.

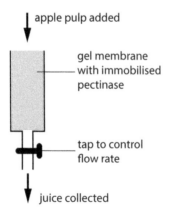

You might write:

'I would make sure that the same mass of apples is added, and that they were the same age.'

This is a basic answer.

A good answer needs to be both accurate and detailed. For example:

'I would make sure that the same mass of apples is added, for example 100 g, and that they were the same age, e.g. 1 week old. I would also control the temperature at an optimum for the pectinases involved, e.g. 30 °C.'

Look at the following example:

A student carried out an experiment to investigate the effect of temperature on respiration in yeast cells. 1 g of dried yeast was added to 25 cm^3 of a 5% glucose solution and after 10 minutes incubation at 15 °C, 1 cm^3 of 5% TTC solution was added. TTC is an artificial hydrogen acceptor which turns from colourless to red in the presence of hydrogen atoms which are released during respiration. The time taken for the yeast solution to turn red was recorded.

The experiment was repeated at 30 °C and 45 °C and the time taken for the yeast suspension to turn red was recorded below.

Temperature (°C)	Time taken for the yeast suspension to turn red (s)			
	Trial 1	Trial 2	Trial 3	Mean (nearest whole second)
15	450	427	466	448
30	322	299	367	329
45	170	99	215	161

Q: What conclusions could be drawn from this experiment regarding the effect of temperature on respiration in yeast?

You might write:

'Increasing temperature decreases the time taken for the yeast suspension to turn red, indicating that respiration is occurring more quickly.'

A good answer needs to be both accurate and detailed, for example,

'Increasing the temperature increases the rate of respiration in the yeast, so dehydrogenase enzymes remove hydrogen atoms from triose phosphate more quickly. This is due to increased kinetic energy of the dehydrogenase enzymes and triose phosphate substrate molecules at higher temperatures. As more hydrogen atoms are released more quickly, so TTC is reduced more quickly, turning the yeast red in a shorter time.'

If asked to comment on the validity of your conclusion, you might write:

'It was difficult to determine when the solutions turned red, making it difficult to know when to stop timing the reactions.'

A good answer would be more detailed. For example:

'The results at 45 °C are very variable and range from 99 to 215 seconds. It is difficult to reach a conclusion about the effect of temperature on respiration in yeast as only three temperatures were investigated. Another major difficulty would be in determining the endpoint of the reaction, as no standard red colour or colorimeter was used.'

As part of this skill, you could also be asked to identify the independent, dependent and controlled variables in an investigation. Remember:

- The independent variable is the one I change.
- The dependent variable is the one I measure.
- Controlled variables are variables which affect the reaction being investigated and must be kept *constant*.

Preparing for the examination

Types of exam question

There are two main types of questions in the exam.

1. Short-answer structured questions

The majority of questions fall into this category. These questions may require description, explanation, application, and/or evaluation, and are generally worth 6–10 marks. Application questions could require you to use your knowledge in an unfamiliar context or to explain experimental data. The questions are broken down into smaller parts, e.g. (a), (b), (c), etc., which can include some 1-mark name or state questions, but most will require description, explanation or evaluation for 2–5 marks. You could also be asked to complete a table, label or draw a diagram, plot a graph, or perform a mathematical calculation.

Some examples requiring 'name', 'state' or 'define':
- Define the term biodiversity. (1 mark)
- State the term used to describe the transfer of energy between consumers. (1 mark)
- Name the cells shown that are undergoing meiosis. (1 mark)
- Identify hormone A shown in the graph. (1 mark)

Some examples requiring mathematical calculation:
- The magnification of the image above is × 32,500. Calculate the actual width of the organelle in micrometres between points A and B. (2 marks)
- Using the graph, calculate the initial rate of reaction for the enzyme. (2 marks)
- Calculate the percentage energy lost through respiration by secondary consumers. (2 marks)
- Use the Hardy–Weinberg formula to estimate the number of individuals in a population of 1000 that would be carriers of the condition. (4 marks)
- Calculate X^2 for the results of the cross shown. (3 marks)

Some examples requiring description:
- Describe how biodiversity loss could be delayed. (1 mark)
- Describe how a sweep net could be used to estimate the diversity index of insects at the base of a hedge. (3 marks)

Some examples requiring explanation:
- Suggest one limitation of the method used and explain how this could have affected the validity of the conclusion drawn. (2 marks)
- Explain why there must be three bases in each codon to assemble the correct amino acid. (2 marks)
- Explain the term planetary boundary. (2 marks)
- Explain why it is important when using a biosensor to measure urea concentration to maintain a constant temperature and pH. (2 marks)
- Explain how the structures of cellulose and chitin are different from that of starch. (2 marks)

Some examples requiring application:
- Suggest the function of NAD in the series of reactions shown. (1 mark)
- A drug has been shown to block the initiation of S phase in mitosis. Suggest why this could be used to treat cancer. (3 marks)
- Use the information provided to explain why sodium benzoate would affect the accuracy of the biosensor. (5 marks)

Some examples requiring evaluation:
- Describe how you could improve your confidence in your conclusion. (2 marks)
- Analyse the data in the table and draw alternative conclusions. Explain how you reached these conclusions. (3 marks)
- Evaluate the strength of their evidence and hence the validity of their conclusion. (4 marks)

2. Extended response questions

One question in each component exam contains an extended response question worth 9 marks. The quality of your extended response (QER) will be assessed in this question. You will be awarded marks based upon a series of descriptors: to gain the top marks it is important to give a full and detailed account, including a detailed explanation. You should use scientific terminology and vocabulary accurately, including accurate spelling and use of grammar and include only relevant information. It is a good idea to do a brief plan before you start, to organise your thoughts: You should cross this out once you have finished. We will look at some examples later.

Command or action words

These tell you what you need to do. Examples include:

Analyse means to examine the structure of data, graphs or information. A good tip is to look for trends and patterns, and maximum and minimum values.

Calculate is to determine the amount of something mathematically. It is really important to show your working (if you don't get the correct answer, you can still pick up marks for your working).

Choose is to select from a range of alternatives.

Compare involves you identifying similarities and differences between two things. It is important when detailing similarities and differences that you discuss both. A good idea is to make two statements, linked with the word 'whereas'.

Complete means to add the required information.

Consider is to review information and make a decision.

Describe means give an account of what something is like. If you have to describe the trend in some data or in a graph then give values.

Discuss involves presenting the key points.

Distinguish involves you identifying differences between two things.

Draw is to produce a diagram of something.

Estimate is to roughly calculate or judge the value of something.

Evaluate involves making a judgment from available data, conclusion or method, and proposing a balanced argument with evidence to support it.

Explain means give an account and use your biological knowledge to give reasons why.

Identify is to recognise something and be able to say what it is.

Justify is about you providing an argument in favour of something; for example, you could be asked if the data support a conclusion. You should then give reasons why the data support the conclusion given.

Label is to provide names or information on a table, diagram or graph.

Outline is to set out the main characteristics.

Name means identify using a recognised technical term. Often a one-word answer.

State means give a brief explanation.

Suggest involves you providing a sensible idea. It is not straight recall, but more about applying your knowledge.

General exam tips

Always read the question carefully: read the question twice! It is easy to provide the wrong answer if you don't give what the question is asking for. All the information provided in the question is there to help you to answer it. The wording has been discussed at length by examiners to ensure that it is as clear as possible.

Look at the number of marks available. A good rule is to make at least one different point for each mark available. So make five different points if you can for a four-mark question to be safe. Make sure that you keep checking that you are actually answering the question that has been asked – it is easy to drift off topic! If a diagram helps, include it: but make sure it is fully annotated.

Timing

There is one written examination paper for each unit, each lasting 2 hours. Each examination is out of 90 marks and contributes 25% of the final grade. In Unit 4, Section B contains a choice of one question from three worth 20 marks: You should ONLY answer the question from the topic that you have studied.

Common exam mistakes

1. **Misreading the question!**
   Sounds obvious I know but READ the question carefully – know your command words!

2. **Not including enough detail**
   You should budget on a minimum of one mark a point, so if the question is out of 5, make at least 5 points, making sure to include your biological knowledge.

3. **Using incorrect terminology**
   Reliability is NOT the same as accuracy!

4. **In maths answers ALWAYS show all your working**
   Credit is given if the examiner can see your steps even if you do end up with the wrong answer, so always show your working in full AND remember your units. If you don't you could lose a mark!

5. **Spelling errors**
   Spelling of key scientific words MUST be right; for example, examiners won't accept 'meitosis' for meiosis because it could be confused with mitosis! The general rule is examiners will allow phonetic spelling so long as it can't be confused for something else. However, for the extended question the quality of your extended response IS assessed.

6. **Describing data**
   Remember to quote data from graphs/tables in your answer WITH units.

7. **Running out of time**
   Exam papers are written to give you plenty of time. If you get stuck, move on BUT remember to come back to it later. Every year marks are lost because some questions are left blank!

8. **Drawing graphs**
   Full marks are rarely awarded for graphs. Common errors include:

   - Incorrect labels on axes
   - Missing units
   - Sloppy plotting of points
   - Failing to join plots accurately
   - Non-linear scales.

   Have a go yourself – can you spot the mistakes?

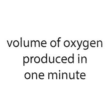
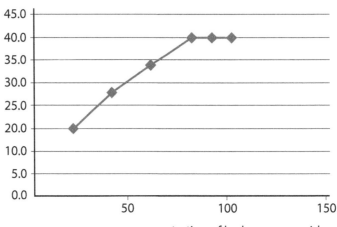

Mistakes are:

- No units on either axis.
- No value for origin on horizontal axis.
- Vertical axis is nonlinear, i.e. gaps are unequal.

Also make sure that you draw range bars and can explain their significance.

Unit 3 Energy, Homeostasis and the Environment

3.1 Importance of ATP and 3.3 Respiration

Topic summary

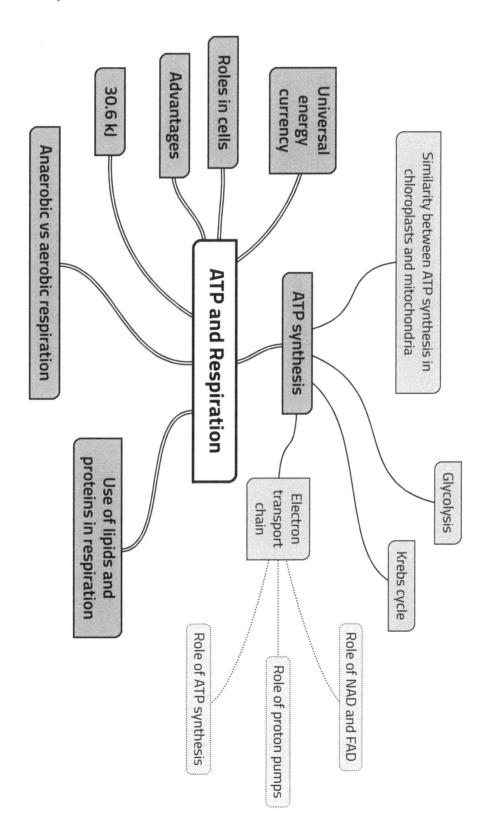

- Roles in cells
- Advantages
- 30.6 kJ
- Anaerobic vs aerobic respiration
- Universal energy currency
- Similarity between ATP synthesis in chloroplasts and mitochondria

ATP and Respiration

- ATP synthesis
- Use of lipids and proteins in respiration
- Electron transport chain
- Glycolysis
- Krebs cycle
- Role of ATP synthesis
- Role of proton pumps
- Role of NAD and FAD

Practice questions

Q1 [AO1]

a) In the space below draw a labelled diagram of ATP. (2)

b) Name **two** uses for ATP in a plant cell. (2)

....

....

....

c) Outline **three** advantages of ATP to a cell. (3)

....

....

....

d) Using examples, distinguish between substrate level phosphorylation and oxidative phosphorylation. (3)

....

....

....

....

....

....

[AO1, AO2]

Q2 Aerobic respiration occurs in a number of stages.

a) Complete the table using a tick (✓) to indicate which statements apply to the following stages in respiration, or a cross (✗) if they do not. (4)

> **EXAM TIP**
>
> Don't leave any boxes blank. If you are unsure have a guess, you have a 50:50 chance of getting it right!

Statement	Glycolysis	Link reaction	Krebs cycle	Electron transport chain
Occurs in the mitochondrial matrix				
ATP produced by substrate-level phosphorylation				
FAD reduced				
NADH$_2$ oxidised				

b) Explain the role of ATP in glycolysis. (3)

....

....

....

....

c) During strenuous exercise muscles may temporarily respire anaerobically. Explain why it is important for the muscles of an athlete to convert pyruvate into lactate (lactic acid). (3)

....

....

....

....

[AO3, AO1]

Q3 An experiment was carried out to investigate the effect of temperature on respiration in yeast. 5 cm³ of yeast suspension was placed into a test tube with 1 cm³ of triphenyl tetrazolium chloride (TTC). TTC is an artificial hydrogen acceptor which is colourless when oxidised but turns red when reduced. Both solutions were equilibrated at each temperature prior to mixing and the experiment carried out three times at each temperature. The results are shown in the graph below:

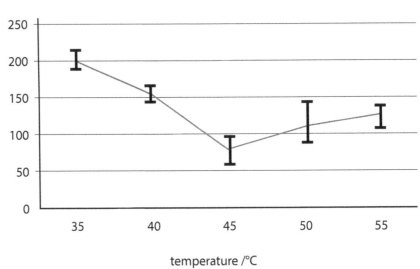

The effect of temperature on respiration in yeast

a) The student concluded the optimum temperature for respiration in the yeast was 45 °C. Evaluate this statement. (3)

....

....

....

....

....

....

b) Explain **two** ways in which the experiment could be improved to increase reliability of the data. (2)

....

....

....

....

c) Explain why the time taken for the TTC to change colour takes longer at 55 °C. (3)

....

....

....

....

....

....

d) Using your knowledge of respiration and the information provided, explain how TTC can be used to measure respiration in yeast. (3)

....

....

....

....

....

....

Unit 3 Energy, Homeostasis and the Environment

Q4

[AO1, AO2, M]

a) Aerobic respiration can yield a theoretical maximum 38 moles of ATP from one mole of glucose. One mole of glucose contains 2880 kJ of energy, and the hydrolysis of ATP liberates 30.6 kJ per mole.

i) State precisely where each step in aerobic respiration takes place in animal cells. (3)

....

....

....

ii) Calculate the energy efficiency of aerobic respiration. Show your working. (2)

Answer ....

b) Anaerobic respiration theoretically yields much less energy, with the efficiency around 2%. Describe anaerobic respiration in animals, explaining why this value is probably much higher. (3)

....

....

....

....

....

[AO1, AO2, S]

Q5 The following diagram shows a stage in aerobic respiration:

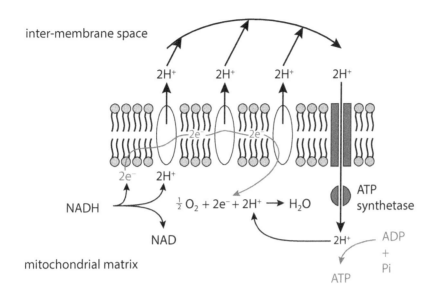

a) What is meant by the term oxidative phosphorylation? (1)

....

....

b) Explain why each NADH generates three ATP molecules, but FADH only generates two. (3)

....

....

....

....

c) Cyanide is a non-competitive inhibitor of the final proton pump in the electron transport chain. Suggest why exposure is fatal and inhibition is not overcome as elections and protons accumulate. (3)

....

....

....

....

[M, AO1, AO2]

Q6 The electron micrograph below shows a mitochondrion from muscle tissue.

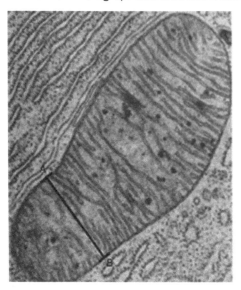

a) Estimate the surface area of the organelle shown using the formula surface area $2\pi r l + 2\pi r^2$, where l = length of organelle 11.2 μm, π = 3.14 and diameter is 1.2 μm. Show your working. (2)

Answer ....

b) The typical surface area of mitochondria found in most cells ranges between 25 and 40 μm². Suggest the advantage of mitochondria with a larger surface area in muscles. (3)

....

....

....

....

[AO1, AO2]

Q7 The diagram below shows the respiratory pathways for carbohydrates, proteins and fats:

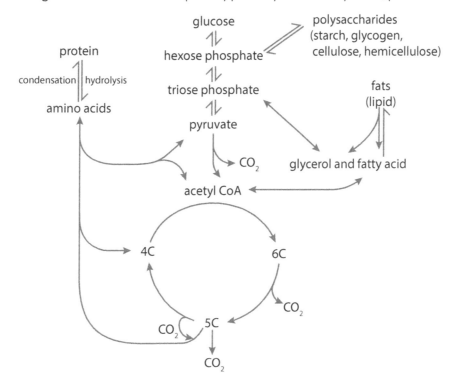

a) On the diagram above, mark where ADP is phosphorylated. (2)

b) Which yields more energy, 1 g of fat or 1 g of carbohydrate? Explain your answer. (2)

....

....

....

....

c) Outline how proteins are respired. (2)

....

....

....

....

Question and mock answer analysis

Q&A 1

[a = AO1, b = AO2, c = AO3]

A sample of calf liver was homogenised in an ice-cold isotonic buffer solution, and then centrifuged at high speed to separate the organelles. The supernatant and a sample of mitochondria were incubated with glucose at 37 °C and the products identified. The experiment was repeated with both samples using pyruvate instead of glucose.

The results are shown in the table:

Sample	Incubated with glucose		Incubated with pyruvate	
	CO_2 produced	Lactate produced	CO_2 produced	Lactate produced
Mitochondria	✗	✗	✓	✗
Supernatant	✗	✓	✗	✗

a) Explain why the buffer used was isotonic and suggest why it was ice-cold. (2)

b) Explain the results observed. (5)

c) What would you expect to see if the experiment were repeated using mitochondrial matrix? Explain your answer. (2).

Lucie's answers

a) Isotonic buffer prevents lysis of mitochondria ✓ and ice-cold slows down enzymes which might damage contents. ✓

b) Mitochondria are unable to metabolise glucose, which is why neither carbon dioxide nor lactate is produced. ✓

> **MARKER NOTE**
> The answer could have included that glycolysis does not occur in the mitochondria and therefore will not have the enzymes necessary to break down glucose.

When mitochondria are incubated with pyruvate, carbon dioxide is produced, because pyruvate can diffuse into the mitochondria and carbon dioxide is produced as a result of Link reaction, and Krebs cycle that occur within the mitochondrial matrix. ✓ The supernatant contains the enzymes present in the cytoplasm of the liver cells and so glycolysis occurs, producing lactate from glucose by anaerobic respiration. ✓ No lactate or carbon dioxide are produced when supernatant is incubated with pyruvate because the link reaction and Krebs cycle do not occur in the cytoplasm. ✓

c) The results would be the same as for the mitochondria ✓ as all the reactions, e.g. link and Krebs that produce carbon dioxide, occur in the matrix. ✓

Lucie achieves 8/9 marks

Ceri's answers

a) Buffer maintains a constant pH. ✗ Ice-cold slows down enzymes. ✓

> **MARKER NOTE**
> Ceri has confused isotonic buffer with pH buffer, and whilst the reason for having the buffer ice-cold could have been more reasoned the idea is there.

b) Mitochondria cannot produce carbon dioxide or lactate. ✗

> **MARKER NOTE**
> This is purely descriptive, there is no explanation.

Mitochondria release carbon dioxide when incubated with pyruvate, because pyruvate is hydrolysed during Link, and Krebs cycle. ✓

> **MARKER NOTE**
> Ceri mentions the reactions but could have included where they take place, i.e. mitochondrial matrix.

The supernatant is made from cell cytoplasm which is where glycolysis occurs, so lactate is produced by anaerobic respiration. ✓

> **MARKER NOTE**
> Ceri did not explain why carbon dioxide and lactate are not produced when the supernatant is incubated with pyruvate, i.e. because the link and Krebs cycle do not occur in the cytoplasm.

c) The results would be the same as for the mitochondria ✓ as all the reactions that produce carbon dioxide occur there. ✗

> **MARKER NOTE**
> Ceri needs to be clearer as to why, i.e. that reactions in link and Krebs cycle that produce carbon dioxide occur in the mitochondrial matrix.

Ceri achieves 4/9 marks

EXAM TIP
It is important that you refer to stages in respiration and state exactly where they occur.

3.2 Photosynthesis

Topic summary

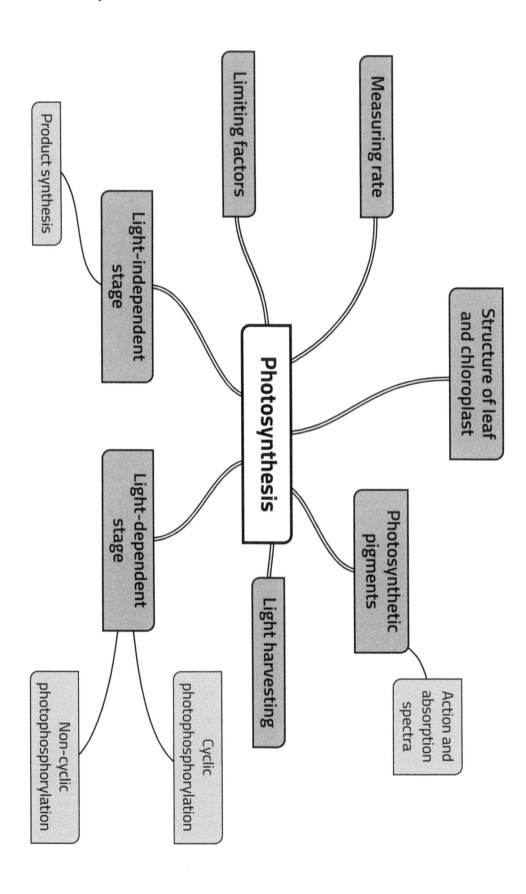

Practice questions

[AO1]

Q1 The diagram below summarises the steps in the light-dependent stage of photosynthesis.

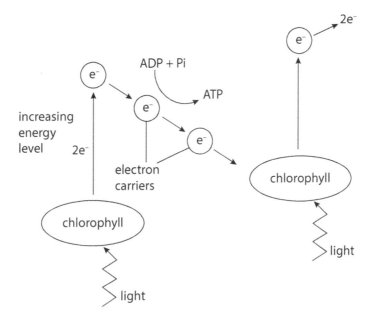

a) State precisely where the stage takes place. (1)

....

b) Name the process by which ATP is produced as shown in the diagram. (1)

....

c) Name the group of biological molecules to which ADP belongs. (1)

....

d) Explain the role of water in the light-dependent stage. (3)

....

....

....

....

e) In the absence of NADP, explain the fate of the electrons. (1)

....

....

....

[AO1, AO2]

Q2 The graph below shows the absorption spectrum for a typical plant:

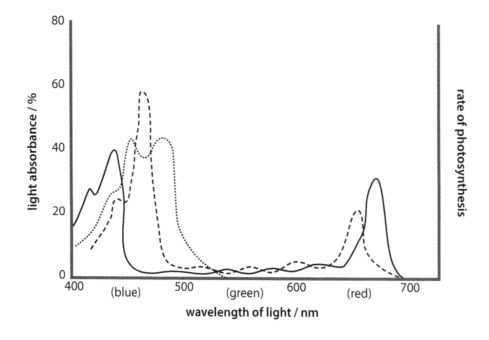

a) Explain why chloroplasts are said to be transducers. (1)

b) Distinguish between absorption spectrum and action spectrum. (1)

c) On the graph, mark which line represents the absorption spectrum for carotenoids. Explain your answer. (2)

d) On the graph draw a line to represent the action spectrum and label it. (1)

e) The wavelength and intensity of light are affected by the depth of water. At 10 m the light intensity is half that at the surface, and comprises only yellow, green and blue wavelengths.

Depth / m	Light intensity as % of light at surface	Percentage of available light by colour / %			
		Red	Yellow	Green	Blue
0	100.0	25	25	25	25
10	50.0	1	33	33	33
20	25.0	0	0	50	50
30	12.5	0	0	0	100

Saccharina latissima (sugar kelp) is a brown-coloured seaweed which has broad fronds to capture light. Sugar kelp is often found at depths of up to 30 m around the coast of the British Isles.

Using the information provided, explain why sugar kelp has large brown fronds and can grow at depths of up to 30 m. (2)

....

....

....

....

[AO3, AO1]

Q3 In 1887 Engelmann conducted an experiment to find the site of photosynthesis. He shone a light through a prism to separate the different wavelengths of light and exposed this to a suspension of algae with evenly distributed, motile, aerobic bacteria. The results are shown in the diagram below:

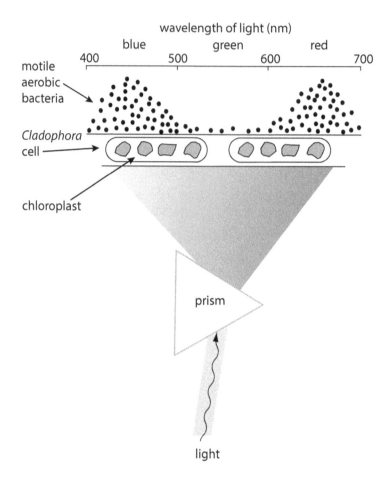

a) Engelmann concluded the *Cladophora* cells only absorbed blue and red wavelengths of light. Using your knowledge and the results from Engelmann's experiment, fully justify this conclusion. (3)

....

....

....

....

b) Outline what is meant by the antenna complex and explain its function. (3)

....

....

....

....

[AO1, AO2]

Q4 The graphs below show the effect of three different limiting factors, X, Y and Z on the rate of photosynthesis in plants:

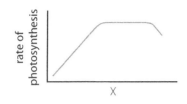 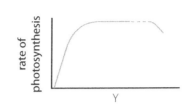 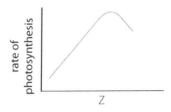

a) Explain fully what is meant by a limiting factor. (1)

....

....

b) Identify the three limiting factors shown. Explain your choice. (6)

X = ...

Reason

....

....

Y = ...

Reason

....

....

Z = ...

Reason

....

....

Q5 [AO1, AO3]

ATP is synthesised in mitochondria and chloroplasts in the same way. Evaluate this statement. (9 QER)

Q6 A photosynthometer with a capillary tube diameter of 0.1 cm was used to measure the volume of oxygen produced by a piece of Canadian pondweed in five minutes at 20 °C. The results are shown in the table below:

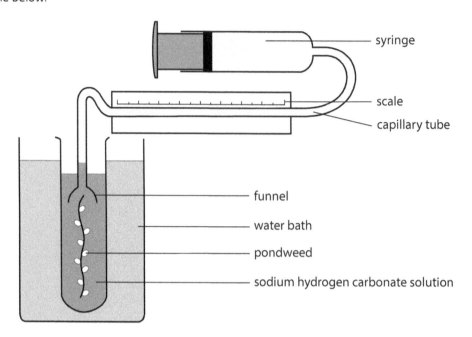

Temperature / °C	Length of bubble in capillary tube / mm				Mean volume of oxygen produced in five minutes / mm³
	Trial 1	Trial 2	Trial 3	Mean	
20	25	23	22		
25	32	34	40		
30	41	42	42		
35	45	47	40		
40	32	30	30		

The volume of the bubble collected is calculated by the formula:

Volume = $\pi\, r^2 \times$ length of bubble

Where π = 3.14

a) Complete the table above. (3)

b) The student concluded the optimum temperature was 35 °C. Evaluate this statement. (3)

....

....

....

....

Question and mock answer analysis

[a = AO1, b = AO2, c = AO3]

Q&A 1 An experiment was carried out using algae in Calvin's lollipop flask. At regular intervals over one hour, samples were removed into a tube which contained hot methanol. The products were identified and their masses measured using mass spectroscopy. The experiment was carried out once using 0.04% hydrogen carbonate and repeated using 0.008%. The relative masses of glycerate-3-phosphate (GP), triose phosphate (TP) and ribulose bisphosphate (RuBP) are shown on the graph below:

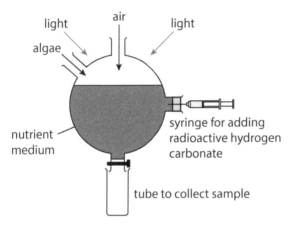
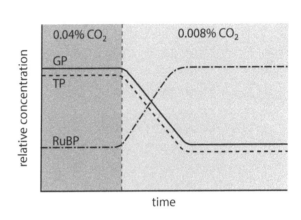

a) Suggest why samples were collected in a tube containing hot methanol. Explain why results would be less reliable if this were not done. (2)

b) Describe and explain the effect of decreasing hydrogen carbonate concentration on the relative concentrations of GP, TP and RuBP. (5)

c) Suggest what the effect of decreasing light intensity would be on the relative concentrations of TP and RuBP with 0.04% hydrogen carbonate. (3)

Lucie's answer

a) Contains hot methanol to denature enzymes to prevent any further reactions ✓ If this was not done further products could be made, e.g. GP could be converted to TP. ✓

b) Decreasing concentration of hydrogen carbonate means that less carbon dioxide is available to join with RuBP to produce GP, so RuBP accumulates. ✓

And therefore less GP can be produced. Any GP present is being converted into TP and so GP falls. ✓ TP concentration falls because of a reducing GP concentration and any TP present will be converted to carbohydrate, so TP is used up. ✓

c) Decreasing light intensity means less ATP and reduced NADP, so less TP is made since ATP and reduced NADP are needed to make TP from GP. ✓ Less RuBP ✓ will be produced because RuBP is still being used up to make GP but RuBP is not being regenerated as GP cannot be made into TP, which is needed to make RuBP. ✓

> **MARKER NOTE**
> Decreasing light intensity means less ATP and reduced NADP, so less RuBP because RuBP is still being used up to make GP but RuBP is not being regenerated as GP cannot be made into TP, which is needed to make RuBP.

Lucie achieves 8/10 marks

Ceri's answer

a) To kill algae to stop reactions. ✗

MARKER NOTE
Ceri fails to say that enzyme reactions will be stopped: This is important as GP could be further converted to TP unless enzymes are denatured.

b) Decreasing concentration of hydrogen carbonate causes RuBP concentration to increase. ✗

MARKER NOTE
This is purely descriptive: There is no link made between hydrogen carbonate and carbon dioxide concentrations, and there is no mention of the enzyme involved RuBisCO or enzyme kinetics.

GP decreases when hydrogen carbonate decreases because GP not being made from RuBP, but is being converted into TP. ✓

MARKER NOTE
Ceri fails to mention the reasons for a decrease in TP: that TP is being converted into carbohydrate.

c) Less TP is made since ATP and reduced NADP are needed to make TP from GP. ✓ Less RuBP will be produced, and more GP will be made. ✓

MARKER NOTE
Ceri correctly predicts what will happen to RuBP and GP but fails to explain why less RuBP is made.

Ceri achieves 3/10 marks

EXAM TIP
It is important that you read the question carefully and give as detailed a response as you can. You must name enzymes involved and explain the results in terms of enzyme kinetics, that were covered in AS. If asked to suggest, an explanation is still required.

3.4 Microbiology

Topic summary

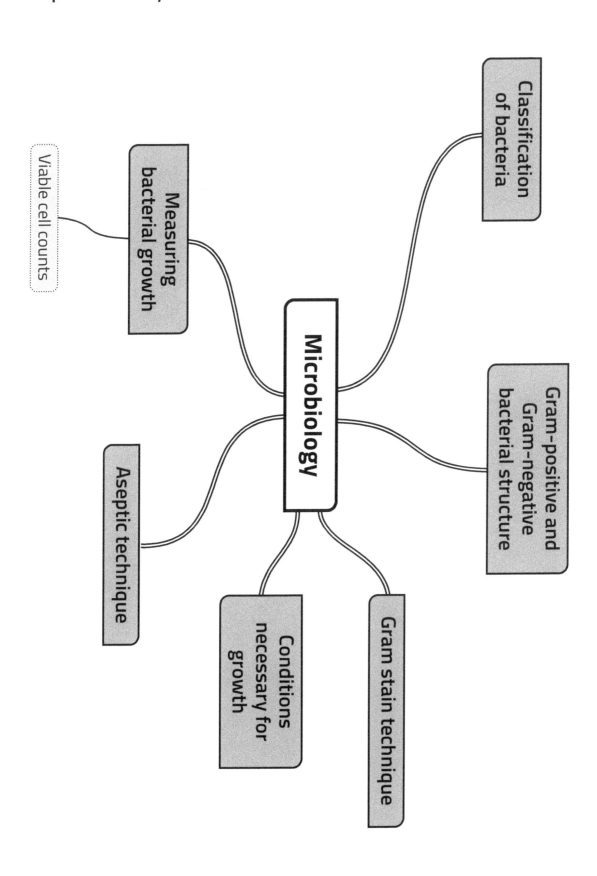

Classification of bacteria

Measuring bacterial growth

Viable cell counts

Microbiology

Gram-positive and Gram-negative bacterial structure

Aseptic technique

Conditions necessary for growth

Gram stain technique

Practice questions

Q1 [AO1, AO2]

Cholera is an infection of the intestines caused by the bacterium *Vibrio cholerae*. The bacteria release toxins that cause watery diarrhoea, leading to severe dehydration and often death. Infection occurs mainly from consuming food or water contaminated by the faeces of an infected person. In the ten months following the earthquake in Haiti in 2010, over 60,000 cases of cholera were reported resulting in 1400 deaths.

a) Suggest how the spread of cholera could have been prevented following the earthquake. (2)

....

....

....

b) The diagram below shows part of the cell wall from *Vibrio cholerae*.

outer layer

Y

i) Name **two** components found in the outer layer. (2)

....

....

ii) Name the component found in layer Y. (1)

....

iii) What colour would you expect the wall of *Vibrio cholerae* to be following the Gram stain? (1)

....

c) Explain why penicillin is an ineffective treatment for cholera. (3)

....

....

....

....

....

d) Describe how scientists could perform a viable cell count to estimate the number of bacteria present in a sample of water. (4)

....

....

....

....

....

....

....

....

Unit 3 Energy, Homeostasis and the Environment

[AO1, AO2, M]

Q2 Three tubes were prepared with nutrient medium and inoculated with three different types of bacteria. They were placed at 30 °C for 24 hours. The distribution of bacteria is shown in the diagrams:

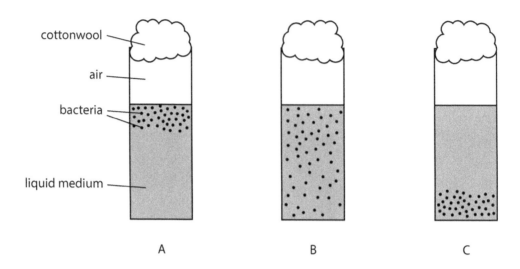

a) Name *two* nutrients other than vitamins or minerals necessary for growth which would need to be added to each tube. Explain your answer. (2)

Nutrient ....

Reason

....

Nutrient ....

Reason

....

b) Suggest which tube contained obligate anaerobes. Give an explanation for your answer. (2)

Tube ....

Reason

....

....

c) Suggest which tube contained obligate aerobes. Give an explanation for your answer. (2)

Tube ....

Reason

....

....

d) Explain what is meant by a facultative anaerobe. (1)

....

....

e) A dilution plate count was performed on the bacteria in tube B to determine the viable cell count by plating 0.1 cm^3 of each dilution. The number of colonies observed are shown in the table below. Calculate the total number of viable cells per cm^3. Show your working. (3)

Dilution factor	Number of colonies seen
10^{-1}	>1000
10^{-2}	>1000
10^{-3}	599
10^{-4}	59
10^{-5}	5

Answer = ...

[AO1, AO2]

Q3 In order to identify bacteria present in a sample of food, a student transferred a sample to a glass slide using aseptic technique.

a) Describe the precautions the student should have taken to ensure the process was carried out aseptically. (2)

....

....

....

b) The student stained the bacteria using the Gram stain, and then viewed under a light microscope. The bacteria were all spherical in shape, but some appeared purple, others pink.

i) Identify the type of bacteria which appeared pink. (2)

....

....

....

ii) Explain why some bacteria stained purple whilst others stained pink. (3)

....

....

....

....

....

....

[AO1, AO2]

Q4

The graph below shows the growth of bacteria over a 24-hour period when exposed to two different carbon sources: glucose and starch.

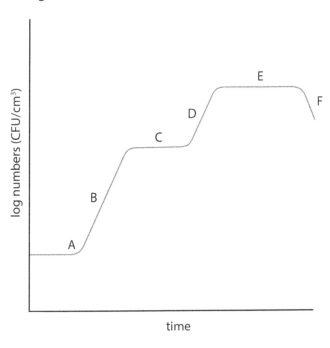

a) Name the different phases A, B, C. (2)

A = ...

B = ...

C = ...

b) Suggest what is happening in phase C which gives rise to phase D. (2)

....

....

....

....

c) Explain what is happening in phase F. (2)

....

....

....

Question and mock answer analysis

[a = AO1, b = AO2, c = AO3, d = AO1]

Following an outbreak of food poisoning, samples of food were tested using the Gram stain, and the bacteria were found to be red in colour. Using the viable cell count method, 1 cm^3 of food sample was diluted by adding to 9 cm^3 of sterile water using aseptic technique. The sample was mixed, and dilutions repeated. 0.1 cm^3 of each dilution was then spread onto a sterile agar plate and the plates incubated at 37 °C for 24 hours. The results are shown below:

Dilution factor	Number of colonies grown
10^{-1}	>1000
10^{-2}	>1000
10^{-3}	899
10^{-4}	81
10^{-5}	7

a) Suggest two reasons why 37 °C was chosen rather than 25 °C to incubate the plates. (2)

b) Identify which dilution factor should be used and calculate the number of live bacteria per cm^3 in the original food sample. (3)

c) Explain why penicillin would not be an appropriate antibiotic to use to treat the patients. (3)

d) Describe how aseptic technique was performed. (3)

Lucie's answers

a) Bacteria would grow faster at 37°C, ✓ and would favour growth of human pathogens. ✓

b) 10^{-4} because 10^{-3} contains too many colonies to count accurately, and 10^{-5} too few. ✓
$81 \times 10,000 \times 10 = 8.1 \times 10^6$. ✓✓

c) The bacteria stained red so must be Gram negative. ✓ Penicillin is ineffective against Gram-negative bacteria. ✓

MARKER NOTE
Lucie would need to include more detail as to why it is ineffective, i.e. because the outer lipopolysaccharide layer prevents the penicillin from reaching the peptidoglycan layer.

d) To ensure aseptic technique was maintained, a sterile pipette was used to transfer the sterile water and food sample, ✓ whilst working close to a bunsen flame. ✓

MARKER NOTE
Lucie could have made reference to using a sterile or flamed spreader.

Lucie achieves 9/11 marks

Ceri's answer

a)   Bacteria grow well at this temperature. ✗

MARKER NOTE

Ceri should include a comparative statement, i.e. that 37 °C would result in faster growth than 25 °C.

b)   81 × 10,000 = 810,000 ✓

MARKER NOTE

Ceri should include an explanation why the plate chosen was used. Some credit can be given for working, but Ceri forgot that only 0.1 cm³ was spread, which dilutes the initial suspension a further ten times.

c)   Penicillin only works against Gram-positive bacteria, which would stain purple, not red. ✓

MARKER NOTE

Ceri needs to include why penicillin is ineffective.

d)   Use a sterile pipette ✓ and spreader. ✓ Keep close to a Bunsen flame. ✓

Ceri achieves 5/11 marks

EXAM TIP

Always show your working in mathematical calculations as some credit may be given even with an incorrect answer.

3.5 Population size and ecosystems

Topic summary

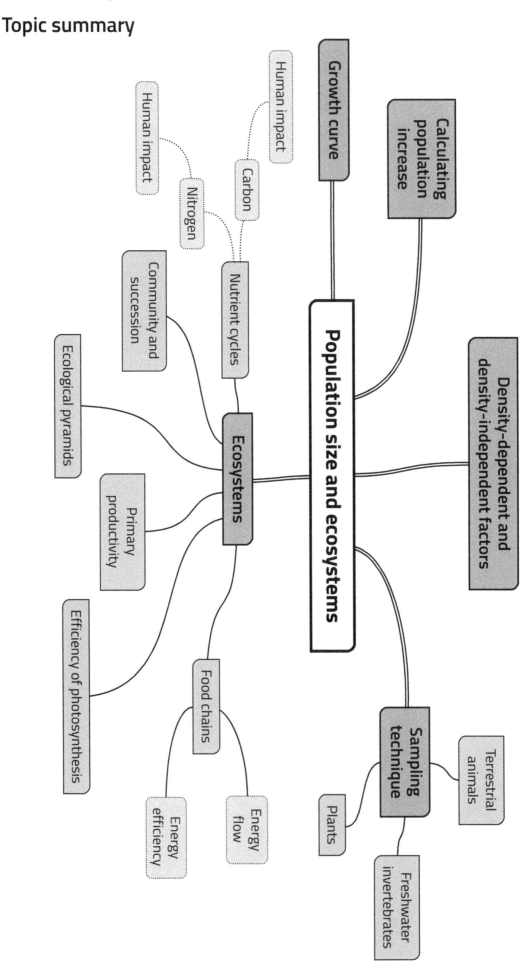

Practice questions

[M, AO1, AO2]

Q1 The graph below shows the growth of bacterial culture over 12 days:

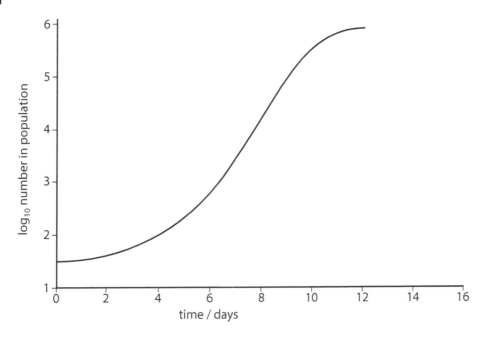

a) On the graph, label the stationary phase. (1)

b) At 12 days, carrying capacity is reached.

    i) Explain what is meant by carrying capacity. (1)

    ....

    ....

    ii) On the graph, draw the growth curve you would expect between 12 and 16 days. (1)

c) Calculate the rate of growth per day between days 4 and 9. Show your working. (2)

    Answer ....

d) Distinguish between density-dependent and density-independent factors, giving an example of each. (2)

    ....

    ....

    ....

Q2 The table below shows the mean net primary production per year in different ecosystems:

Ecosystem	Mean net primary production g m^{-2} year^{-1}
Tropical reef	2450
Tropical rainforest	2250
Estuaries	1450
Temperate deciduous forest	1250
Cultivated land	625
Tundra and alpine	125
Desert	90

a) Define ecosystem. (2)

....

....

....

b) Explain why not all the net primary production is available to the next trophic level. (1)

....

....

....

c) Explain why the mean net primary production differs between the tropical rainforest and deciduous rainforest. (3)

....

....

....

....

....

[M, AO2, AO1]

Q3 The following diagram shows the energy flow through a typical woodland ecosystem:

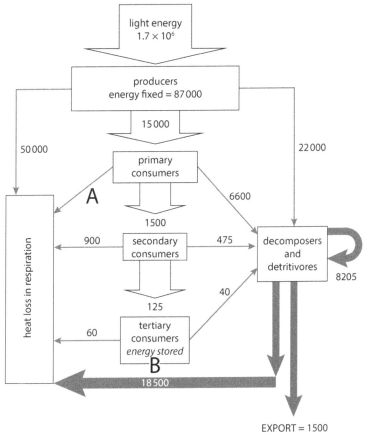

All values in kJ m^{-2} yr^{-1}

a) Calculate the efficiency of the producers. Show your working. (2)

Answer ....

b) Explain why the efficiency of tertiary consumers is much higher than that of producers. (2)

....

....

....

c) State *three* reasons why only a fraction of the sun's energy is fixed by plants. (2)

....

....

....

Q4

[AO1]

Describe how the composition of a community can change over time, from bare rock to a climax woodland, explaining factors which may affect the process. [9 QER]

[AO1]

[M, AO2]

Q5 Fish excrete ammonia as nitrogenous waste from the deamination of excess amino acids in their diets. High levels of ammonia and nitrite are toxic to fish. When establishing a new Koi carp fishpond in your garden it is important to introduce fish slowly. It is recommended to regularly test the water to establish levels of ammonium, nitrite and nitrate ions during this process.

a) The concentration of ammonium, nitrite and nitrate ions was measured in a pond during the first 30 days of its establishment. The results are shown below:

Time / days	Concentration of nitrogen/ mg dm^{-3}		
	Ammonium ions	Nitrite	Nitrate
0	0	0	0
3	4	0	0
6	7	1	0
9	3	7	1
12	1	10	4
15	1	8	9
18	1	4	16
21	1	1	12
24	1	1	8
27	0	0	6
30	0	0	4

Draw a graph to show how concentration of ammonium, nitrite and nitrate changed. (5)

b) Suggest reasons for the change in ammonium ions seen between day 6 and 12. (3)

....

....

....

....

c) Describe and explain the change seen in nitrite levels between day 6 and 21. (2)

....

....

....

....

d) When were aquatic plants introduced? Explain your answer. (3)

....

....

....

....

[AO2]

Q6   The diagram below shows part of the carbon cycle:

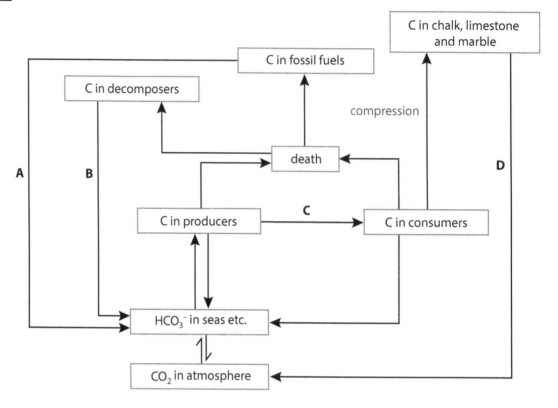

a)   Identify processes A, B, C and D. (3)

A = ...

B = ...

C = ...

D = ...

b)   Explain how carbon dioxide is stored in ocean rocks. (3)

....

....

....

....

....

....

Question and mock answer analysis

Q&A 1

[M, AO1]

The diagram shows the energy flow through a woodland ecosystem. The photosynthetic efficiency per year, represents the proportion of light energy that is available to plants which is converted (fixed) into chemical energy.

a) Use the information provided to calculate the photosynthetic efficiency of the producers expressed as a percentage to 2 decimal places. (2)

b) Calculate the energy lost from primary consumers to decomposers and detritivores (A) per day, to 2 decimal places. (2)

c) Distinguish between gross and net primary productivity. (2)

d) Calculate and explain why the efficiency of tertiary consumers is much higher than primary or secondary consumers. (3)

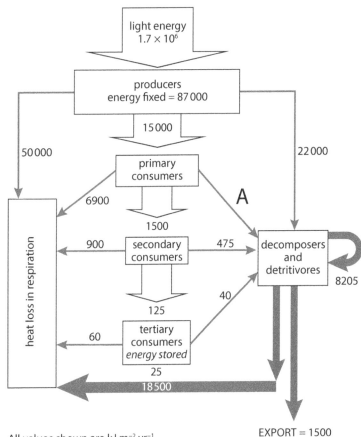

All values shown are kJ m^{-2} yr^{-1}

Lucie's answer

a) $\dfrac{87,000}{1,700,000} \times 100$ ✓ $= 5.12\%$. ✓

b) $15,000 - 6900 - 1500 = 6,600$ ✓

$\dfrac{6,600}{365} = 18.08 \ kJ \ m^{-2} \ day^{-1}$. ✓

c) GPP represents the rate at which producers convert light energy into chemical energy ✓ whereas NPP represents the rate at which energy is converted into biomass which is available to the next trophic level. ✓

d) Efficiency of primary consumers is 10%, and secondary consumers is 8.3%, which is much lower than tertiary consumers which is 20% ✓ This is because tertiary consumers are more efficient at digesting their protein-rich food. ✓

> **MARKER NOTE**
> It is also due to the fact there is very little energy left at the top of the food chain.

Lucie achieves 8/9 marks

Ceri's answer

a) $\dfrac{87,000}{1,700,000} \times 100$ ✓ $= 5.11\%.$ ✗

MARKER NOTE

Ceri forgot to round to 2 decimal places, so only 1 mark awarded for method.

b) $15,000 - 6900 - 1500 = 6,600$ ✓

MARKER NOTE

Correct calculation per year, but needed to divide by 365 to calculate per day, and include units.

c) $GPP = NPP - R$ ✗

MARKER NOTE

This is the correct equation but a comparison between the two terms should be included.

d) Tertiary consumers' efficiency is twice that of other consumers ✓ and is due to them being more efficient at digesting food.

MARKER NOTE

Ceri needs to say why, i.e. that there is very little energy left at the top of the food chain and so they have to be more efficient at digesting their protein-rich diets.

Ceri achieves 3/9 marks

EXAM TIP

It is important that you read the question carefully. Make sure that you follow all steps in any calculation.

3.6 Human impact on the environment

Topic summary

Unit 3 Energy, Homeostasis and the Environment

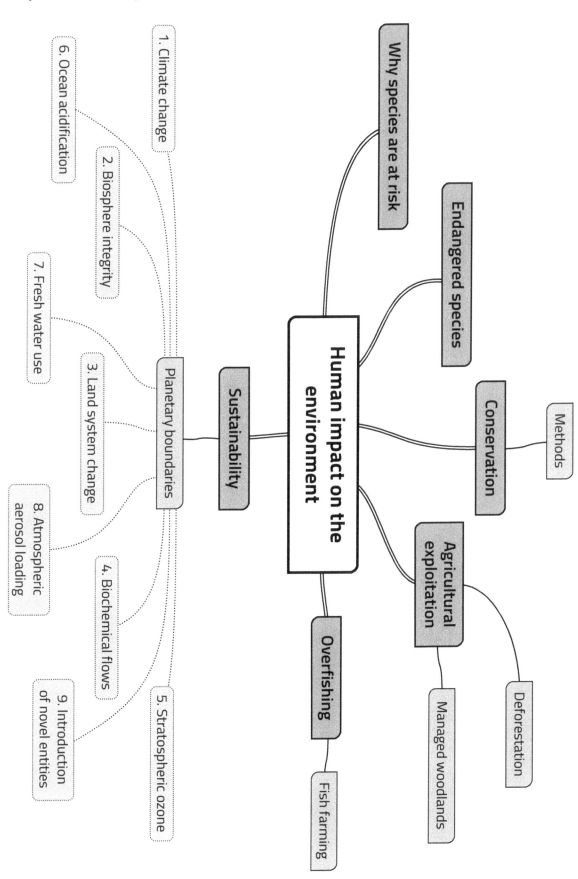

Practice questions

[AO1, AO2]

Q1 The Bornean orangutan (*Pongo pygmaeus*) is now critically endangered. Bornean orangutans live only on the island of Borneo, where their populations have declined by 60% since 1950. The maps show forest cover in Borneo since 1950, and the graphs show orangutan numbers and palm oil production over a similar period.

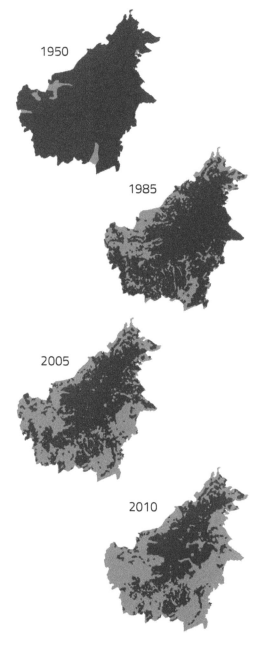

1950

1985

2005

2010

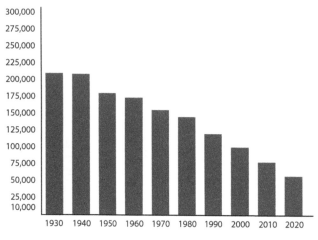

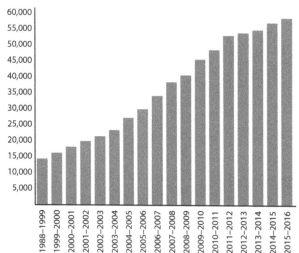

Using the information and your knowledge, explain the factors involved in the decline in orangutan numbers in Borneo, and strategies that could be used to reverse the decline. [9 QER]

[AO1, AO2]

Q2 Nine earth system processes and their boundaries have been identified which mark the safe zone for the planet. The use of biofuels has reduced the net carbon emissions being released into the atmosphere.

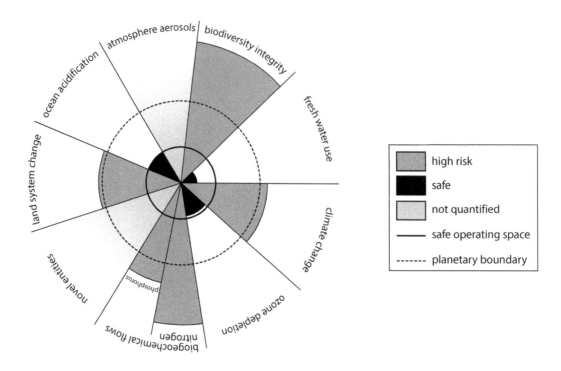

a) Explain why biofuels reduce carbon emissions but are not completely carbon-neutral. (2)

....

....

....

....

b) Explain how other planetary boundaries have been impacted as a result of this change. (2)

....

....

....

....

c) In years to come, climate change is likely to impact on the availability of fresh water. Describe *three* ways in which the availability of fresh water can be increased. (2)

....

....

....

....

Question and mock answer analysis

[a = AO1/AO2, b = M, c & d = AO2]

Q&A 1 The following graph shows atmospheric carbon dioxide levels up to the year 2000:

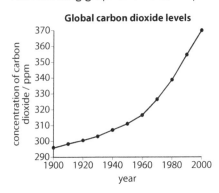

Global carbon dioxide levels

Scientists have determined that the planetary boundary for climate change, which is determined by atmospheric carbon dioxide, is 350 ppm.

a) What is meant by a planetary boundary? Using information in the graph, explain why the climate change boundary has been crossed. (3)

b) Calculate the average rate of increase in atmospheric carbon dioxide between 1920 and 2000 and use this to estimate the atmospheric carbon dioxide concentration in 2030. (3)

c) Explain why the value you calculated in part b) is likely to be inaccurate, and how you could improve the accuracy of your calculation. (3)

d) Biofuels are increasingly being used to try to reduce carbon emissions as they are regarded as carbon-neutral. Explain why biofuels like bioethanol produced from sugar cane are not truly carbon-neutral and why they impact other planetary boundaries. (3)

Lucie's answer

a) It is a framework proposed by environmental scientists that marks a safe zone for humanity as a precondition for sustainable development. ✓

The planetary boundary for climate change as represented by atmospheric carbon dioxide was crossed in the late 1980s. ✓

> **MARKER NOTE**
> Lucie's answer could have been expanded to include the reasons, e.g. increased combustion of fossil fuels, deforestation, mechanised agriculture, use of fertilisers.

b) 370 – 300 = 70 ppm ✓

$\frac{70}{80}$ = 0.875 ppm yr^{-1}

0.875 × 30 = 26.3 ✓

26.3 + 370 = 396.3 ppm. ✓

c) The CO_2 concentration in 2015 already exceeded this. ✓ I calculated the value using an average rate of increase between 1920 and 2000, but the rate of increase in CO_2 increased in the 1960s so my value will be lower. ✓ It would be better to calculate the rate from 1980–2000 and use this. ✓

d) Energy is used in their production, processing and distribution. ✓ Switching land use to grow sugar cane results in species loss, which impacts biodiversity integrity boundary further. ✓

> **MARKER NOTE**
> Land system change boundary is also impacted.

Lucie achieves 10/12 marks

Ceri's answer

a) They are proposed by scientists to mark a safe zone for the planet.

MARKER NOTE
Ceri should include reference to sustainability.

The current atmospheric carbon dioxide is higher than 350 ppm. ✗

MARKER NOTE
Information from the graph, e.g. that the value in 2000 was 370, or that 350 ppm was crossed in late 1980s should be included.

b) 385 ppm. ✗

MARKER NOTE
It looks like Ceri miscalculated the increase by incorrectly reading off the graph. Whilst Ceri has added 15 ppm to 370 ppm, without any working it is not possible to give any credit.

c) The CO_2 concentration in 2015 is 400 ppm which already exceeds this. ✓

MARKER NOTE
Ceri needs to explain where the inaccuracy may have come from, i.e. that the rate of increase changed during the period, or how it could be improved.

d) Energy is needed to make and transport the biofuels ✓

MARKER NOTE
This is correct but Ceri does not say which planetary boundaries are impacted.

By deforesting an area to grow sugar cane, many valuable species are lost

Ceri achieves 2/12 marks

EXAM TIP
Learn your definitions! It is important to show working in any calculations, so credit for method can be given. Identifying inaccuracies is a difficult skill, so look carefully for any anomalies and try to identify any assumptions that have been made.

[AO1]

Q&A 2 Explain the importance of different farming activities used to maximise efficient food production, and the consequences upon planetary boundaries of their use. (9 QER)

Unit 3 Energy, Homeostasis and the Environment

Lucie's answer

Farmers need to maximise plant growth in order to maximise food production. Plants need a source of nitrogen to synthesise proteins needed for growth and to make enzymes and hormones. Growth rates are increased in soils that have a good supply of nitrogen. ✓ Plants obtain nitrogen from the soil, usually in the form of nitrates. In the soil, bacteria recycle nitrogen through the nitrogen cycle. Ploughing and drainage are both important because they aerate the soil. This is important because oxygen is needed for the active transport of mineral ions including nitrates into the plant roots. ✓

MARKER NOTE

Lucie could have included the use of manure and its breakdown through ammonification to increase nitrogen content.

It also promotes nitrification where Nitrosomonas bacteria convert ammonium ions into nitrites, and Nitrobacter bacteria which convert nitrites into nitrates. ✓ Both of these bacteria respire aerobically and so need oxygen to do this. The process of denitrification, which is carried out in the soil by Pseudomonas bacteria converting nitrates back into atmospheric nitrogen, is an anaerobic process, and so is inhibited in well-aerated soils. Where soils lack nitrogen, farmers can plant leguminous plants such as peas and clover. ✓ These plants have root nodules which contain Rhizobium which are nitrogen-fixing bacteria that are able to increase soil nitrogen content when they are ploughed back into the soil.

Farmers also can apply nitrogen-based fertilisers like ammonium nitrate, which is produced by the Haber process. ✓ This requires much energy from fossil fuels to make them, which causes atmospheric pollution in the form of carbon dioxide, which is a greenhouse gas. Activities such as ploughing involve the use of machinery, which also causes carbon dioxide pollution. These activities have led to increases in carbon dioxide emissions, which has meant that the climate change boundary has been crossed. ✓

MARKER NOTE

Lucie could have included a definition of what is meant by a planetary boundary. This then would put into context why exceeding the boundaries is an important consequence.

Farming and the removal of hedgerows to allow for ever larger machinery have led to the extinction of species, as habitats such as hedgerows have been lost. This activity, and other habitat losses have led to the biodiversity boundary being exceeded. ✓ The excessive removal of atmospheric nitrogen during the Haber process has resulted in the biochemical boundary for nitrogen being crossed. ✓

EXAMINER COMMENTARY

Lucie gives a full and detailed account of the different farming activities and how they influence the nitrogen cycle. The effect on three planetary boundaries has been discussed. The account is articulate and shows sequential reasoning. There are no significant omissions.

Lucie achieves 8/9 marks

Ceri's answer

Plants need nitrogen to grow. It is needed to manufacture proteins.

MARKER NOTE
Ceri needs to explain in more detail why proteins are needed, e.g. to synthesise enzymes.

Plants take up nitrogen in the form of nitrates from the soil. Farmers can do a lot to help increase the amount of nitrogen in the soil in order to maximise growth of crops, for example they can add fertilisers and manure. ✓ There are bacteria in the soil that help to break down organic waste into nitrates by a process called nitrification, e.g. Nitrosomonas and Nitrobacter.

MARKER NOTE
This lacks detail. The answer does not explain that Nitrosomonas bacteria convert ammonium ions into nitrites, and Nitrobacter bacteria convert nitrites into nitrates.

Ploughing also helps as it mixes manure through the soil, improving soil texture and improves oxygenation of soil.

MARKER NOTE
There is no mention of drainage, or why increased oxygenation improves soil nitrogen, i.e. that nitrification is an aerobic process and that nitrate uptake by roots requires oxygen for active transport. The use of leguminous plants and nitrogen fixation is omitted.

A soil rich in oxygen also inhibits denitrification where nitrates are converted back into atmospheric nitrogen by bacteria called Pseudomonas denitrificans. ✓ The overuse of inorganic fertilisers has affected a number of planetary boundaries, e.g. Biochemical boundary for nitrogen which has been exceeded. ✓

MARKER NOTE
There is no mention of climate change boundary being exceeded or why.

Monoculture has also affected the biodiversity boundary.

MARKER NOTE
There is no detail on why and how the biodiversity boundary has been exceeded, e.g. due to habitat loss from hedgerow removal.

EXAMINER COMMENTARY
Ceri gives a limited account of the different farming activities and how they influence soil fertility. Two planetary boundaries have been discussed but the effect on the biochemical boundary is not mentioned. Ceri makes some relevant points, correctly names three bacterial species involved in the nitrogen cycle but could have detailed nitrification more clearly. There is limited use of scientific vocabulary.

Ceri achieves 3/9 marks

EXAM TIP
Remember it is not a mark per point, but rather what you say and how you say it. Answers must not omit any key information and should include all key scientific terminology to gain top marks. Watch your spelling too!

3.7 Homeostasis and the kidney

Topic summary

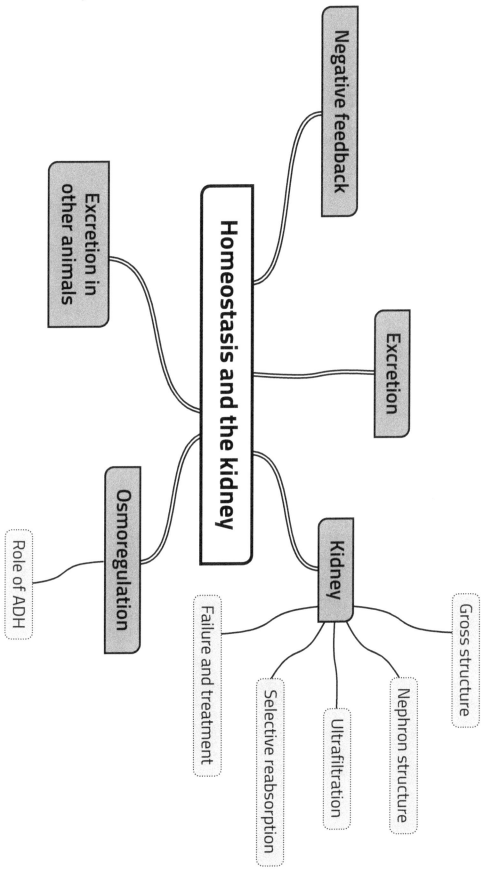

Practice questions

[AO1, S]

Q1 The following diagram shows a nephron in a mammalian kidney:

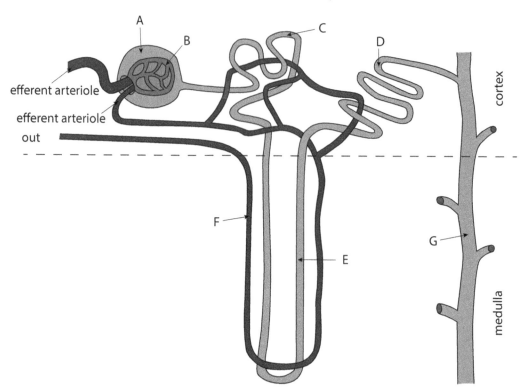

a) Complete the table below, some have been done for you. (7)

Letter	Name	Function
A		
B		
C		
D		Control of blood pH
E		
F	Vasa recta	
G		

b) Name the part of the kidney where E, F and G are found. (1)

c)   State what is meant by water potential. Explain how part E affects the water potential of urine. (6)

....

....

....

....

....

....

....

....

....

....

d)   Explain why part E is longer in desert animals. (3)

....

....

....

....

e)   Explain how ADH affects the permeability of the collecting duct. (3)

....

....

....

....

Unit 3 Energy, Homeostasis and the Environment

[AO2, AO1]

Q2 The following photograph shows a high-power light microscope image of a TS from part of the mammalian kidney:

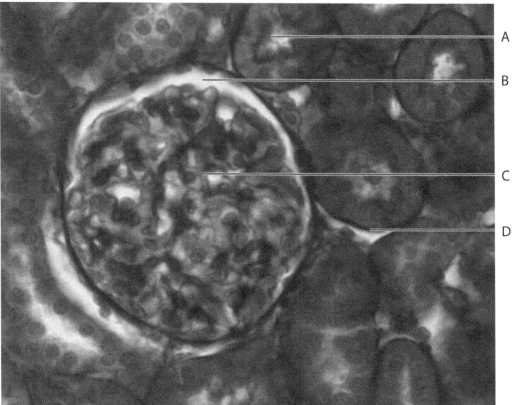

a) Identify the **four** structures labelled A–D visible in the photograph. (3)

A ....

B ....

C ....

D ....

b) Name the part of the kidney from which the specimen was taken. (1)

....

c) State the function of structure A. Explain **two** features which adapt structure A for its function. (3)

Function ....

....

....

....

....

[AO1, AO2, M]

Q3 A section of the kidney as seen under the light microscope is shown below. The image is ×400 magnification.

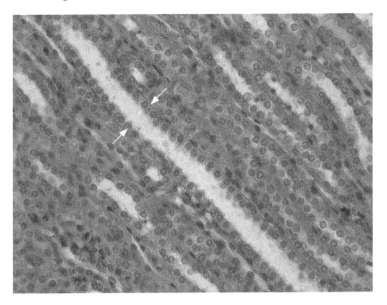

a) Name the part of the kidney from which the specimen was taken, explaining the reason for your choice. (2)

Part of kidney ....

Reason

....

....

....

b) Calculate the width of the structure shown by the two arrows. Show your working. (2)

Answer ....

c) State the function of the membrane lining the structure shown. How is structure adapted to perform its function? (2)

....

....

....

Q4

[AO1]

Describe and explain how the different regions of the nephron are adapted to their functions. [9 QER]

Q5

[AO1]

Different organisms excrete different nitrogenous materials. In its simplest form ammonia is excreted by freshwater fish, whilst birds, insects and reptiles expend much energy to convert ammonia to uric acid.

a) Describe how ammonia is produced in fish. (2)

....

....

....

b) Explain why freshwater fish are able to excrete ammonia whilst mammals are not. (2)

....

....

....

c) Suggest why birds, reptiles and insects expend much energy to convert ammonia to uric acid. (3)

....

....

....

....

....

[AO1, AO2]

Q6 Using your knowledge of the functioning of the kidney, answer the following questions.

a) Human chorionic gonadotropin (hCG) is a hormone which can be detected in the urine of pregnant women. Suggest why hCG is present in the urine of pregnant women. (2)

....

....

....

....

b) Explain why glucose can be present in the urine of diabetic patients, but not healthy individuals. (2)

....

....

....

....

c) Explain why the concentration of urea in the Bowman's capsule is approximately 0.35 g dm^{-3}, rising to over 6 g dm^{-3} in the collecting duct. (2)

....

....

....

....

Question and mock answer analysis

[a & b = AO1, c = AO2, d = AO3]

Q&A 1 The table below shows the typical concentrations of two solutes (glucose and urea) in three different regions of the kidney nephron, labelled P, R and S, in the diagram below.

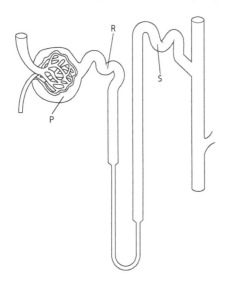

Solute	Mean concentration / g dm^{-3}		
	Bowman's capsule	Proximal convoluted tubule	Distal convoluted tubule
Glucose	0.12	0.00	0.00
Urea	0.35	0.65	6.25
Sodium ions	0.28	0.24	0.02

a) State exactly where you would expect to find the structure labelled P in a cross-section through the kidney. (1)

b) Explain how the changes in concentration for glucose and urea are brought about. (5)

c) Suggest why diabetic patients can suffer from damage to cells lining the distal convoluted tubule. (3)

d) Scientists have concluded the majority of sodium ions are reabsorbed after the proximal convoluted tubule. Explain the evidence in the data which supports this conclusion. (2)

Lucie's answer

a) In the cortex. ✓

b) The mean concentration of glucose decreases from 0.12 g dm^{-3} in region P to 0.00 g dm^{-3} in regions R and S ✓ because glucose is selectively reabsorbed into the blood in region R. ✓ Urea is not selectively reabsorbed in this region, but water is, ✓ so the concentration of urea increases from 0.35 g dm^{-3} to 6.25 g dm^{-3} because the same mass of urea is dissolved in a smaller volume of water. ✓

> **MARKER NOTE**
> Lucie should have included details of the mechanism of reabsorption here, i.e. glucose is reabsorbed by secondary active transport with sodium ions and water by osmosis.

c) High levels of blood glucose will mean not all glucose can be reabsorbed in the proximal convoluted tubule ✓. The remaining glucose will lower the water potential of the filtrate.

> **MARKER NOTE**
> No link to cellular damage has been made. Lucie could have included that lower water potential would cause cells lining tubule to crenate.

d) The mean concentration of sodium ions entering proximal convoluted tubule is between 0.24 and 0.28 g dm^{-3} ✓ In the distal convoluted tubule, the concentration has fallen to 0.02 d dm^{-3} indicating most have been reabsorbed ✓

Lucie achieves 8/11 marks

Ceri's answer

a) Cortex. ✓

b) Glucose is reabsorbed by co-transport with sodium ions ✓ causing its concentration to decrease. ✓

MARKER NOTE
It is important to quote data when explaining.

Urea is not reabsorbed in regions R and S ✓ which is why its concentration increases. ✗

MARKER NOTE
The reason given is incorrect: Water is selectively reabsorbed by osmosis in the proximal convoluted tubule and the loop of Henle, so the same mass of urea is dissolved in a smaller volume of water causing its concentration to increase.

c) Glucose will still be present in the distal convoluted tubule which will damage cells ✗

MARKER NOTE
Ceri has not made the link between glucose lowering water potential and the loss of water from cells causing them to crenate.

d) Sodium ion concentration decreases between the proximal convoluted tubule and the distal convoluted tubule, but some ions do remain. ✓

MARKER NOTE
No reference to data is made, but the answer does indicate some ions do remain.

Ceri achieves 5/11 marks

EXAM TIP
Explain answers fully and always quote data to support your answer. This is a good example where knowledge of osmosis from AS is directly relevant to A2, and will often be chosen by examiners to be tested.

Unit 3 Energy, Homeostasis and the Environment

3.8 The nervous system

Topic summary

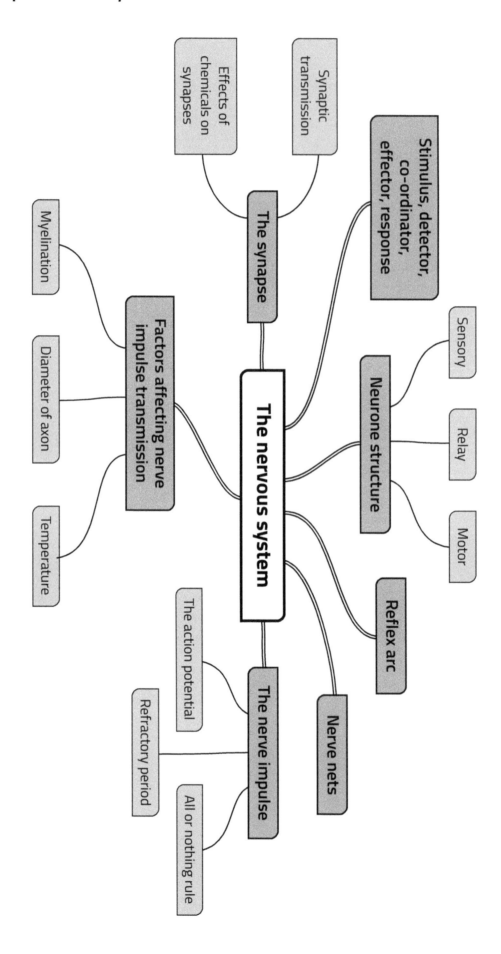

Practice questions

Unit 3 Energy, Homeostasis and the Environment

[AO2, AO1, S]

The drawing below shows a mammalian neurone.

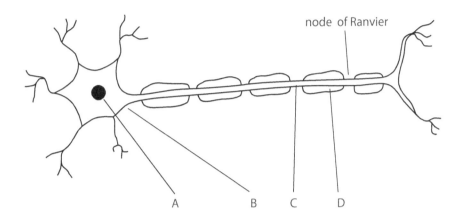

a) Name the type of neurone shown. (1)

....

b) Draw an arrow to show the direction of impulse. (1)

c) Identify parts A–D labelled. (3)

A ....

B ....

C ....

D ....

d) Explain **two** factors which increase the speed of impulse conduction in this neurone. (2)

....

....

....

....

....

[AO2, AO1]

Q2 The action potential in a motor neurone was measured by inserting a microelectrode into the axon and a reference electrode placed in the extracellular fluid outside the axon. The voltage measured across the axon during the passage of an impulse is shown below:

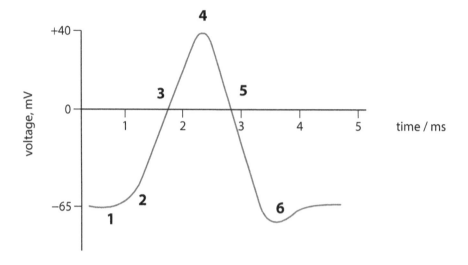

a) Name the state that exists at point 1. (1)

....

b) Describe the events that occur to maintain the state at point 1. (3)

....

....

....

....

....

c) Mark the relative refractory period on the graph above. (1)

d) On the graph, draw an arrow to the show direction of the impulse. (1)

e) Distinguish between absolute and relative refractory periods. (2)

....

....

....

....

f) Describe the events which occur to bring about the change in voltage seen between points 1 and 4 on the graph. (5)

....

....

....

....

....

....

[AO1, AO2, AO3]

Q3 The diagram below shows a synapse:

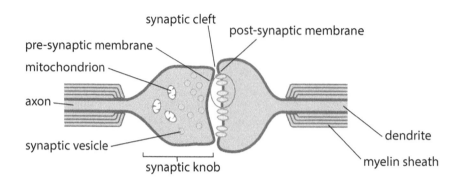

a) Explain the role of calcium ions in the functioning of the synapse. (3)

....

....

....

....

....

b) An experiment was carried out to measure the effects of amphetamines on synaptic transmission. Doses of 0, 1 and 4 mg/kg of methamphetamine sulphate were given to lab rats and the concentration of synaptic acetylcholine measured over a three-hour period. The results are shown in the graph below. Using the information provided and your knowledge answer the following questions.

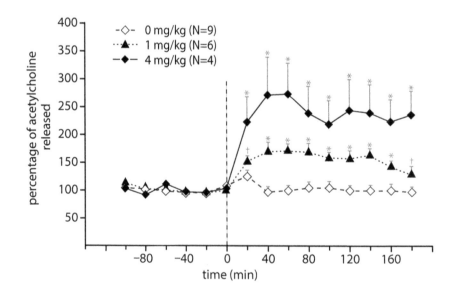

i) Suggest what group of drugs methamphetamine belongs to. Explain your answer. (3)

Group ....

Reason

....

....

....

ii) How confident are you in the results shown? (3)

....

....

....

....

....

c) The experiment was repeated with 0, 0.1 and 0.4 mg/kg of nicotine and the results are shown below:

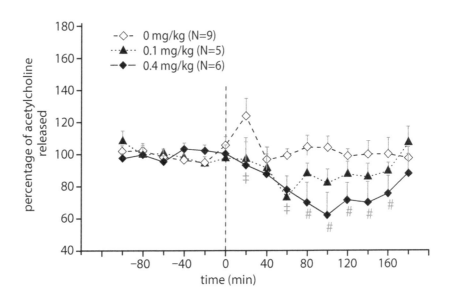

Nicotine mimics the effects of acetylcholine in mammals. Using the results, explain why people can become addicted to nicotine. (2)

....

....

....

....

Q4

[AO1]

Describe synaptic transmission and explain the effect of organophosphorus insecticides on synapses. [9 QER]

[AO2, AO1]

Q5 The diagram below shows a reflex arc:

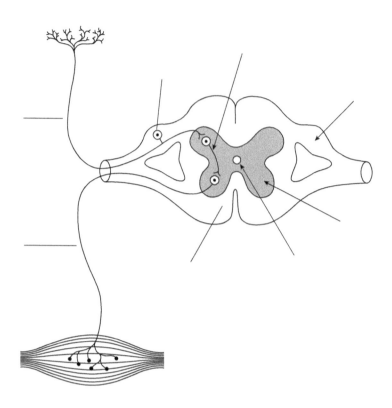

a) On the diagram, label the following structures. (4)

   i) Relay neurone

   ii) Dorsal root ganglion

   iii) White matter

   iv) Motor neurone

   v) Cell body

....

b) Describe the nerve pathways involved in the reflex response occurring when your hand touches a hot surface. (5)

....
....
....
....
....
....

Question and mock answer analysis

Q&A 1

[AO2]

Parkinson's disease is a progressive neurological condition resulting from the death of brain cells that produce dopamine, the neurotransmitter involved in motor control pathways in the brain. Patients are unable to control fine motor movements such as walking. Treatment involves the use of L-dopa which is a synthetic drug that is converted to dopamine in the brain.

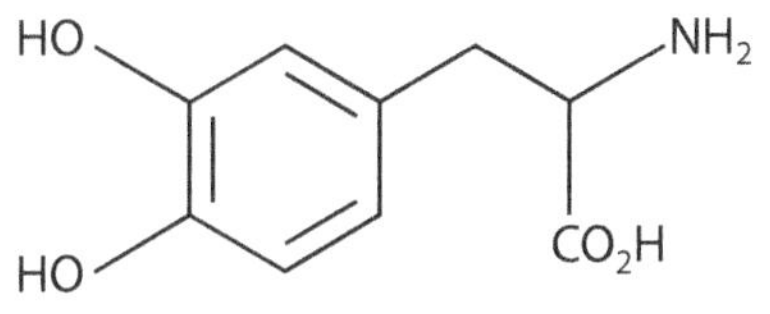

L-dopa

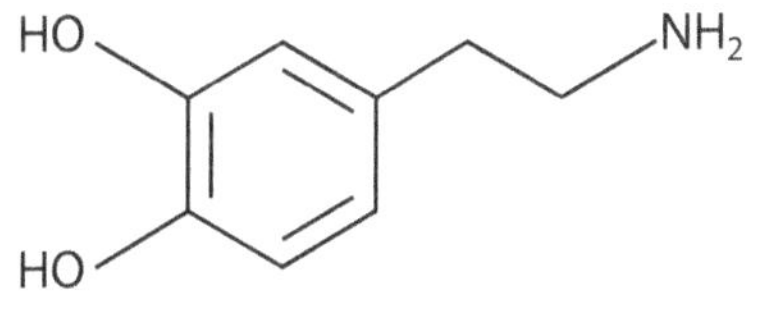

Dopamine

Suggest how L-dopa treats sufferers of Parkinson's disease. (3)

Lucie's answer

Dopamine is a neurotransmitter involved in synapses in the brain responsible for fine motor control. Because less neurotransmitter is released, fewer post-synaptic neurones will be depolarised, which will cause fewer muscles fibres to contract, resulting in difficulty in walking. ✓ L-dopa is decarboxylated ✓ into dopamine by a one-step reaction in the brain, supplying dopamine quickly to the affected areas. The increase in neurotransmitter allows more post-synaptic neurones to be depolarised, so more muscles fibres contract, making walking easier. ✓

Lucie achieves 3/3 marks

Ceri's answer

People who suffer from Parkinson's disease, don't produce enough dopamine because brain cells that produce it have died. This results in poor motor control because fewer motor neurones will be depolarised. ✓

> **MARKER NOTE**
> Need to link to muscle contractions – that fewer muscle fibres will contract.

L-dopa works as a precursor to dopamine and is easily converted into dopamine in the brain.

> **MARKER NOTE**
> Ceri should use information in the diagram, i.e. the loss of carbon dioxide, which is decarboxylation.

The increased levels of dopamine restore function allowing more motor neurones to be depolarised, which makes walking easier. ✓

Ceri achieves 2/3 marks

EXAM TIP

It is important here to use the information in the diagrams and make the link that decarboxylation, i.e. removal of CO_2, will produce dopamine.

[AO1, AO2]

The diagram below shows a section through a motor neurone:

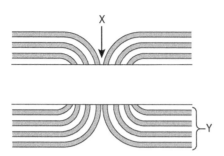

a) Identify X and Y. (1)

b) In multiple sclerosis, the protective myelin coat surrounding nerve fibres is destroyed, a process known as demyelination. Use your knowledge of nerve impulse transmission to explain why this results in poorer reflexes. (3)

Lucie's answer

a) X= node of Ranvier
Y = myelin sheath ✓

b) Loss of myelin results in slower nerve impulse transmission because ions can no longer only pass across the membrane at the nodes of Ranvier where there is no myelin, ✓ so depolarisation no longer only occurs at these nodes, it occurs across the entire length of the neurone ✓ The impulse no longer jumps from node to node. ✗

MARKER NOTE
The action potential 'jumps' from node to node, not the impulse.

Lucie achieves 3/4 marks

Ceri's answer

a) X= axon ✗

Y = myelin sheath

MARKER NOTE
Y is correct but both X and Y needed for the mark.

b) The loss of the sheath slows down impulse transmission because depolarisation occurs at these gaps in the sheath rather than at each node of Ranvier. ✓

MARKER NOTE
Ceri fails to mention the myelin prevents influx of ions, or action potential.

Ceri achieves 1/4 marks

Unit 4 Variation, Inheritance and Options

4.1 Sexual reproduction in humans

Topic summary

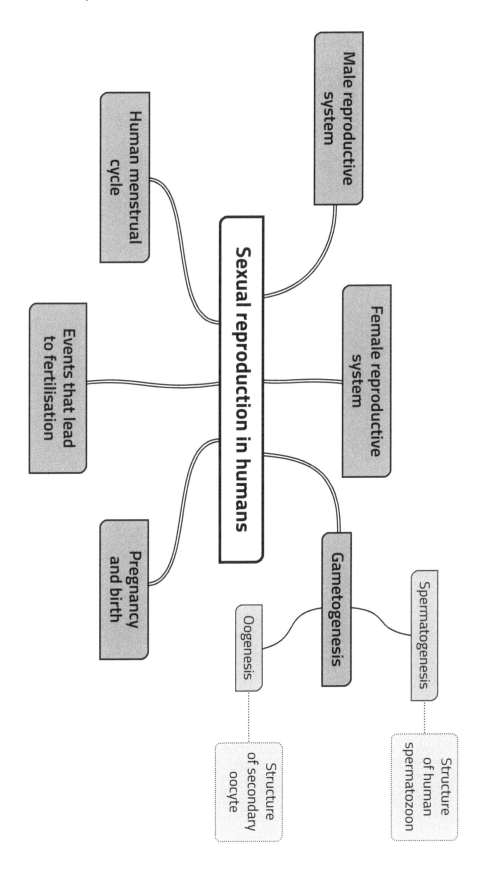

Practice questions

[AO2, AO1]

Q1 The following diagram shows a section through a mammalian testis:

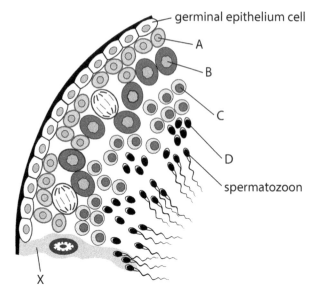

a) Identify cells B, C and D. (2)

B = ...

C = ...

D = ...

b) Explain how cell C differs from cell B. (2)

....

....

....

c) Describe how cell D changes to enable it to fertilise a secondary oocyte. (4)

....

....

....

....

....

....

[AO2, AO1]

Q2   The following diagram shows the male reproductive system:

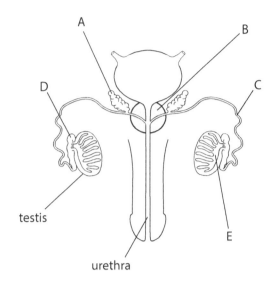

a)   Identify structures A–E. (3)

A = ...

B = ...

C = ...

D = ...

E = ...

b)   State the function of A and B. (2)

A = ...

B = ...

c)   In the space below draw a fully annotated diagram of a human spermatozoon. (3)

Q3

[AO2, AO1]

The following graph shows the levels of four different hormones during the menstrual cycle. Using the graph and your knowledge, answer the following questions.

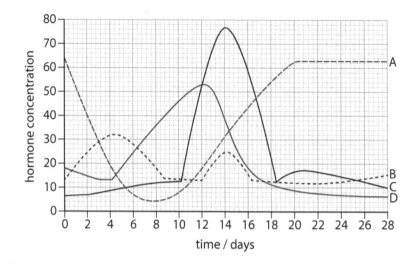

a)   Name hormones A–D. Suggest a reason for your choice. (8)

A = ...

Reason

....

....

B = ...

Reason

....

....

C = ...

Reason

....

....

D = ...

Reason

....

....

b) State precisely where hormone B is released from. (1)

....

c) Use the graph to explain which ovarian hormone could be injected to *stimulate* ovulation. (2)

....

....

....

....

d) Use the graph to explain which ovarian hormone could be injected to *prevent* ovulation. (2)

....

....

....

....

Q4 [AO1]

a) Distinguish between acrosome reaction and cortical reaction. (2)

....

....

....

....

b) In the space below draw a fully annotated diagram of a secondary oocyte. (3)

c) Describe three protective functions of the placenta during pregnancy. (3)

....

....

....

....

....

[AO2, AO1]

Q5 The following drawing represents a human ovary:

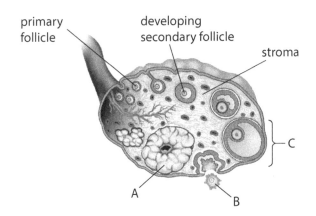

a) Identify structures A–C. (2)

A = ...

B = ...

C = ...

b) Distinguish between the production of the secondary oocyte and secondary spermatocyte. (4)

....

....

....

....

....

....

Unit 4 Variation, Inheritance and Options

Question and mock answer analysis

Q&A 1 [a = AO2, b = AO2/AO1, c = AO1]

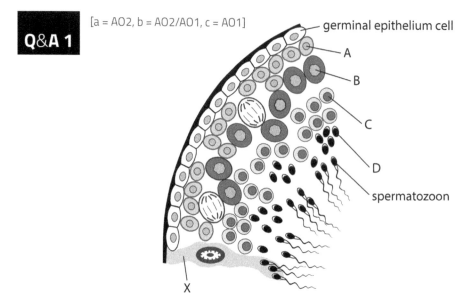

a) Identify cell X and describe its function. (2)

b) Identify cell A and describe the process by which cell B is produced. (2)

c) Explain how the spermatozoon is adapted for its function. (3)

Lucie's answer

a) Cell X is the Sertoli cell. ✓ It provides nutrition for the spermatozoa. ✓

b) Cell A is spermatogonium. ✓ It undergoes mitosis to produce the diploid primary spermatocyte. ✓

> **MARKER NOTE**
> Good, concise answer.

c) The acrosome in the head of the spermatozoa contains hydrolytic enzymes which allows it to digest the zona pellucida of the ovum. ✓

> **MARKER NOTE**
> Prior to the sperm entering zona pellucida, the female gamete is referred to as the secondary oocyte, as meiosis II hasn't occurred yet.

A mid piece is added containing numerous mitochondria and a tail which provides movement toward the secondary oocyte. ✓

> **MARKER NOTE**
> Lucie should include the role of the numerous mitochondria, i.e. to provide the ATP necessary for locomotion.

Lucie achieves 6/7 marks

Ceri's answer

a) X is the nurse cell. ✗ It protects the sperm. ✗

MARKER NOTE

Nurse cell is incorrect, only Sertoli cell is accepted. Protection unqualified is too vague.

b) A is the spermatogonium, ✓ it undergoes meitosis to produce cell B. ✗

MARKER NOTE

Meitosis could be confused with meiosis so cannot be allowed.

c) The spermatozoa has a head containing an acrosome, a mid piece containing mitochondria and a tail for movement. ✓

MARKER NOTE

Ceri needs to explain the function of the parts of the spermatozoa enabling fertilisation of the secondary oocyte, i.e. acrosome contains hydrolytic enzymes which digest the zona pellucida and mitochondria that produce ATP for movement.

Ceri achieves 2/7 marks

EXAM TIP

Watch your spelling with key words. Phonetic spelling is allowed so long as it cannot be confused with another term. Link structure to function, i.e. explain the function.

Unit 4 Variation, Inheritance and Options

4.2 Sexual reproduction in plants

Topic summary

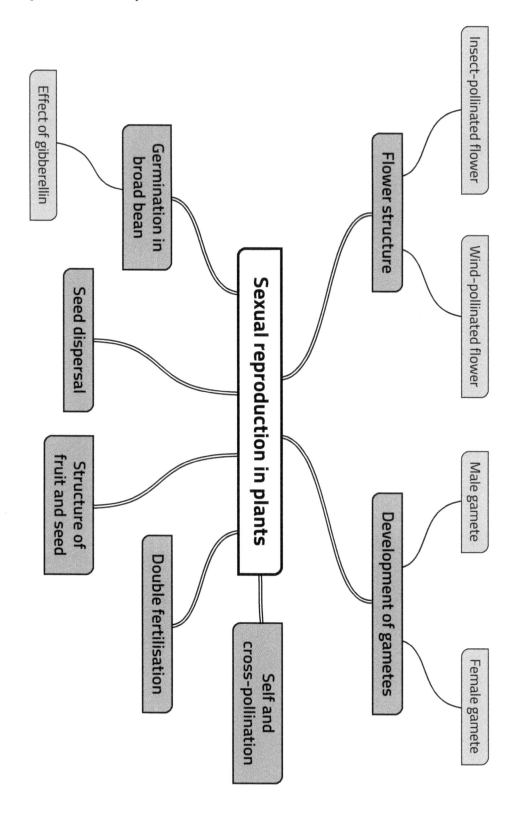

Practice questions

[AO2, AO1]

Q1 The following drawings are of insect-pollinated and wind-pollinated flowers:

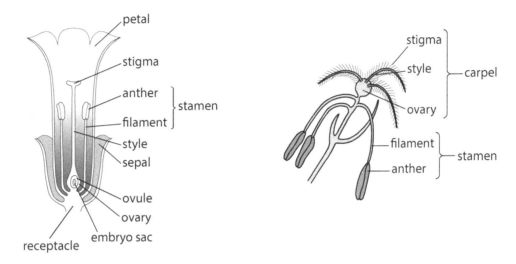

a) Distinguish between the reproductive structures of the two flowers shown in the diagrams, explaining how they are adapted for their mode of pollination. (3)

b) Explain how pollen produced in each flower is adapted to the mode of pollination. (2)

[AO1, AO2, AO3]

Q2 The diagram below shows the events which follow pollination in an insect-pollinated flower:

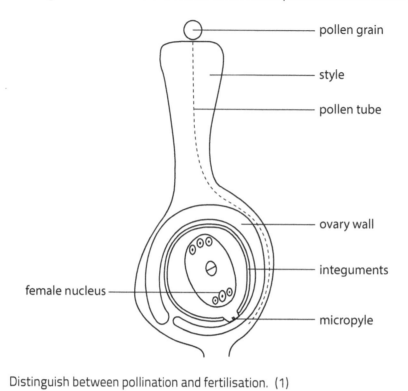

- pollen grain
- style
- pollen tube
- ovary wall
- integuments
- micropyle

female nucleus

a) Distinguish between pollination and fertilisation. (1)

....

....

....

b) Describe the events which occur after pollination leading to fertilisation. (3)

....

....

....

....

....

....

c) Pollen grains from one species will not germinate on the stigma of another species. This is largely thought to be controlled by the sucrose concentration on the stigma. An experiment was carried out to investigate the effect of different sucrose concentrations on pollen germination and pollen tube length. Ten pollen grains were transferred to a glass microscope slide and a drop of sucrose solution added. The slides were then placed into a Petri dish containing a filter paper soaked in water and the lid secured. The slides were left for 24 hours and then viewed under a light microscope at high power (×100) and percentage germination and pollen tube length measured. The experiment was repeated using a range of sucrose concentrations from 0.0 to 1.6 mol dm^{-1}. The results are shown below:

Sucrose concentration/mol dm^{-1}	% germination	Mean pollen tube length/μm
0.0	10	90
0.2	50	200
0.4	70	290
0.8	20	190
1.6	0	0

i) Suggest why the slides were placed into a sealed Petri dish with damp filter paper. (1)

....

....

ii) Suggest why no coverslip was applied to the glass slide. (1)

....

....

iii) The student concluded the optimum sucrose concentration was 0.4 mol dm^{-1}. Evaluate this statement. (3)

....

....

....

....

[M, AO2, AO3]

Q3 An experiment was performed to investigate the effect on gibberellic acid on stem elongation in pea plants (*Pisum sativum*). Ten days following germination, two seedlings of the same height were taken and two drops of a gibberellic acid solution (10 mg of gibberellic acid per 100 cm³ of distilled water) were applied to the apical meristem of each seedling. The experiment was repeated with distilled water as a control. The height of the pea seedlings was measured every two days until harvesting at 20 days. The results are shown below:

Time following germination/days	Mean height of pea plants grown in water/mm	Mean height of pea plants grown in 10 mg of gibberellic acid/mm
10	6	6
12	7	10
14	11	16
16	14	25
18	18	31
20	22	38

a) Calculate the mean percentage increase in seedling height at 20 days. Show your working. (2)

Answer ....

b) Using the results and your knowledge of the effects of gibberellic acid on germination, suggest a conclusion which can be drawn from the results. (2)

....

....

....

....

c) Evaluate your conclusion. (2)

....

....

....

Question and mock answer analysis

Q&A 1

[AO3, AO1]

An experiment was carried out to investigate the optimum requirements for germinating broad bean seeds. Ten seeds were placed in the different environments shown in the table below and the mean seedling height was recorded ten days after germination.

Temperature / °C	Volume of water supplied per day/cm³	Mean seedling height / mm
10	15	41
10	30	48
10	60	9
20	15	65
20	30	72
20	60	13
30	15	82
30	30	86
30	60	18

Using the results, draw a conclusion as to the optimum conditions required for seedling growth in the broad bean. Using your biological knowledge, explain your conclusion. [9 QER]

Lucie's answer

The results show that both temperature and water affect growth in germinating broad bean seedlings. The optimum conditions for growth are 30 °C and 30 cm³ of water supplied, which produced a mean seedling height of 86 mm after ten days. ✓ Too much water inhibited growth, which was most noticeable at 10 °C where the mean seedling height for 60 cm³ was just 9 mm. ✓

> **MARKER NOTE**
> Data is quoted, but some could have been processed, e.g. % increase.

Water is needed for germination and seedling growth. Water is absorbed by the seed causing the testa to split as the tissues swell. The radicle emerges and begins to absorb more water. Water is important because it mobilises enzymes and provides water for the hydrolysis of starch into maltose. ✓

> **MARKER NOTE**
> Hydrolysis of maltose is mentioned, but Lucie could have included an equation or talked more about the chemical addition of water breaking the glycosidic bonds. It would be better to say that hydrolysis of maltose is followed by respiration of glucose. Reference to oxygen being needed for aerobic respiration to fuel biosynthesis for growth could be included.

Once the seedling is photosynthesising, water is also needed for photosynthesis and transport of sucrose to the growing points. ✓ An optimum temperature is important not only for enzymes because temperature increases the kinetic energy of enzyme and substrate molecules resulting in more enzyme–substrate complexes, but for the enzymes involved in photosynthesis, e.g. RuBisCo. ✓ Too much water inhibited growth, for example, even at the optimum of 30 °C, increasing the water supplied from 30 to 60 cm³ caused a reduction in height of seedling from 86 to 18 mm. ✓ Water is needed for growth, but too much reduces the oxygen available to the roots of the developing seedling, as water replaces air in the air spaces found within the soil. Oxygen is needed for aerobic respiration of maltose within the germinating seed and later for the active uptake of mineral ions, e.g. nitrates into the roots. Nitrates are essential for growth because they are needed by plants to synthesise proteins, e.g. enzymes and structural proteins. ✓

> **EXAMINER COMMENTARY**
> Lucie gives a full conclusion which is supported by detailed biological knowledge from both AS and A2. The account is articulate and shows sequential reasoning. There are no significant omissions.

Lucie achieves 7/9 marks

Ceri's answer

The best conditions for growth were 30 °C and 30 cm³ of water supplied. ✓

> **MARKER NOTE**
> Data must be quoted, e.g. 86 mm height, and the conclusion needs to be more detailed.

Water is important because it is needed to hydrolyse starch into maltose within the seed and is needed for photosynthesis.

> **MARKER NOTE**
> Ceri should include details of hydrolysis of maltose and why water is needed for photosynthesis, and translocation of solutes, etc.

30 °C provides the best growth because temperature increases the kinetic energy of enzyme and substrate molecules resulting in more enzyme–substrate complexes. ✓

> **MARKER NOTE**
> Ceri should name some enzymes, e.g. maltase, RuBisCO.

Above 30 cm³ height of seedling was reduced, perhaps because too much water reduces the oxygen in the soil.

> **MARKER NOTE**
> Ceri should include why oxygen is important to the growing seedling, e.g. active uptake of nitrates and other mineral ions.

Oxygen is needed for aerobic respiration in the seed. Aerobic respiration yields more ATP than anaerobic respiration.

> **MARKER NOTE**
> Ceri gives a brief conclusion and explanation of the results. Ceri makes some relevant points, but there is limited use of scientific vocabulary.

Ceri achieves 2/9 marks

> **EXAM TIP**
> Be specific and name both enzyme and substrate. Always include data to support your conclusion.

4.3 Inheritance

Topic summary

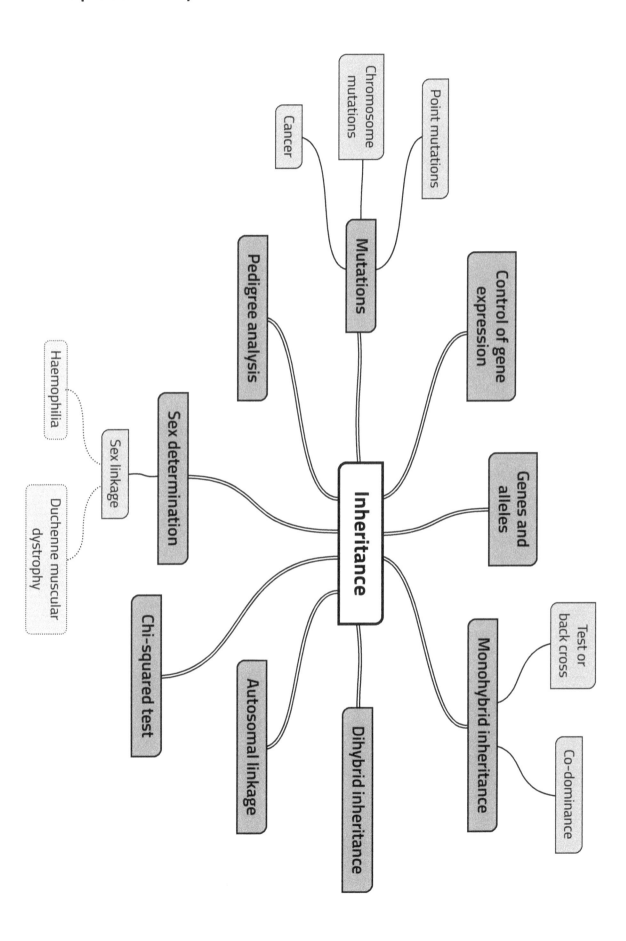

Inheritance

- Mutations
  - Chromosome mutations
  - Point mutations
  - Cancer
- Pedigree analysis
- Control of gene expression
- Genes and alleles
- Monohybrid inheritance
  - Test or back cross
  - Co-dominance
- Dihybrid inheritance
- Autosomal linkage
- Chi-squared test
- Sex determination
  - Sex linkage
    - Haemophilia
    - Duchenne muscular dystrophy

Unit 4 Variation, Inheritance and Options

Practice questions

Q1

[AO1, S]

Distinguish between the following terms:

a) Gene and allele. (2)

b) Homozygous and heterozygous. (2)

c) Dominant and recessive. (2)

d) Sex linkage and autosomal linkage. (2)

Q2

[AO2, AO1]

a) A yellow wrinkled pea (heterozygous for colour and texture) was crossed with a homozygous recessive green smooth pea. Perform a genetic cross and determine the genotypes and phenotypes of the offspring expected. (5)

Parental genotype   X   ....

Gametes

Phenotype ratio   ....

b) State Mendel's first law: The law of segregation. (2)

....

....

....

Q3 [AO2, AO1]

a) Show the results of a genetic cross including the phenotype ratio of the resulting offspring between two heterozygous pea plants with yellow coloured and round seeds, where the recessive phenotypes are green and wrinkled seeds. (5)

Parental genotype   X   ....

Gametes

Phenotype ratio   ....

b) Explain how these results support Mendel's second law of independent assortment. (2)

....

....

....

c) For the cross in part a) show the results of the cross if both genes for colour and seed texture were subject to autosomal linkage. (2)

Parental genotype   X   ....

Gametes

Phenotype ratio ....

d) In the example above in part c) explain two ways in which a greater number of combinations of offspring could arise. (4)

....

....

....

....

....

Unit 4 Variation, Inheritance and Options

[M, AO2]

Q4 In guinea pigs the allele for black coat is dominant to albino and the allele for rough coat is dominant to smooth coat. A heterozygous black smooth-coated guinea pig is mated with an albino guinea pig with smooth coat.

a) Perform a genetic cross to show the result of this cross. (5)

Parental genotype .... X ....

Gametes

Phenotype ratio ....

b) In the first generation, the offspring had the following phenotypes: 27 black rough coat; 22 black smooth coat; 28 albino rough coat; 23 albino smooth coat. Use χ^2 to find out if there is a significant difference between the observed and expected numbers of offspring of the different phenotypes. (6)

Chi-squared table

Degrees of freedom	P = 0.10	P = 0.05	P = 0.02
1	2.71	3.84	5.41
2	4.61	5.99	7.82
3	6.25	**7.82**	9.84
4	7.78	9.49	11.67
5	9.24	11.07	13.39

Using the formula $\chi^2 = \dfrac{\Sigma (O-E)^2}{E}$

Show your working.

Category	Observed (O)	Expected (E)			
black rough coat					
black smooth coat					
albino rough coat					
albino smooth coat					
	$\Sigma =$				$\Sigma =$

$\chi^2 =$ ....

Answer

....

....

....

Unit 4 Variation, Inheritance and Options

[AO2, M]

Q5 The diagram shows the inheritance of haemophilia, a sex-linked condition, within a family.

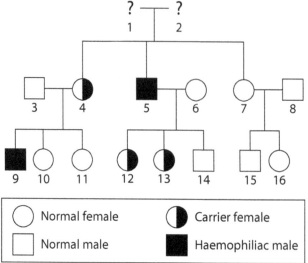

a) Name the term given to the genetic diagram shown above. (1)

....

b) Work out the phenotype of parents 1 and 2. Explain your reasons for each parent. (3)

Phenotype = ...

Reasons

....

....

....

c) Perform a cross to show the chances that another male child from parents 3 and 4 will have haemophilia. (4)

Chance = ...

Q6

[AO2, M]

DMD is a fatal neuromuscular disorder caused by a recessive mutation on the X chromosome. DMD affects about 1 in 3,500 males; however, most girls born with DMD mutations are only carriers, so although not themselves affected by DMD, they can pass DMD genes on to their children. The following diagram shows the inheritance of DMD in a family:

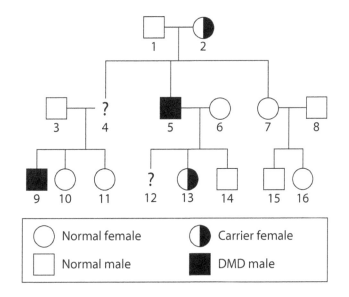

◯ Normal female	◖ Carrier female	
▢ Normal male	■ DMD male	

a) Explain why boys are affected, but girls are carriers. (1)

....

....

b) State the phenotype of parent 4. Explain your choice. (2)

Phenotype = ...

Reasons

....

....

....

c) Perform a genetic cross to calculate the chances of female child 12 being a carrier. (4)

Chance = ...

Question and mock answer analysis

[M, AO2]

Male Anole lizards court females by bobbing their heads up and down while displaying a colourful throat patch. Females prefer to mate with males with a red throat patch who bob their heads fast. These traits are dominant to males with yellow patches and who bob their heads slowly. A male lizard heterozygous for fast head bobbing and red throat patch mated with a female slow head bobbing and yellow throat patch. In the first generation the offspring (F1) had the following phenotypes: 27 fast bobbing red patch; 22 fast bobbing yellow patch; 28 slow bobbing red patch; 23 slow bobbing yellow patch.

a) Complete the cross below to show how the offspring in the first generation inherited the phenotype shown above. (5)

Parental genotype .... X ....

Gametes .... X ....

F1 genotypes

F2 phenotypes

Phenotype ratio

b) Using the table below calculate χ^2 for the results of the cross. (3)

Category	Observed (O)	Expected (E)			

Using the formula $\chi^2 = \dfrac{\Sigma (O-E)^2}{E}$

χ^2 = ...

c) Use your calculated X^2 value and the probability table to conclude how coat colour and texture are inherited. (4)

Degrees of freedom	p = 0.10	p = 0.05	p = 0.02
1	2.71	3.84	5.41
2	4.61	5.99	7.82
3	6.25	7.82	9.84
4	7.78	9.49	11.67
5	9.24	11.07	13.39

Lucie's answer

a) Ff Rr x ffrr ✓

FR, Fr, fR, fr x ffrr ✓

      fr

FR   FfRr   fast bobbing, red patch

Fr   Ffrr   fast bobbing, yellow patch ✓

fR   ffRr   slow bobbing, red patch

fr   ffrr   slow bobbing, yellow patch ✓

ratio is 1:1:1:1 ✓

b)

Category	Observed (O)	Expected (E)	$O - E$	$(O - E)^2$	$(O - E)^2/E$
fast bobbing red patch	27	25	2	4	0.16
fast bobbing yellow patch	22	25	−3	9	0.36
slow bobbing red patch	28	25	3	9	0.36
slow bobbing yellow patch	23	25	−2	4	0.16
Σ	100	100			1.04

✓ (under Expected)   ✓ (under last column)

$X^2 = 1.04$ ✓

c) The null hypothesis is that there is no significant difference between the observed and expected values. ✓

Because the calculated value 1.04 is less than the critical value at p = 0.05, 7.82, we can accept the null hypothesis, so any differences between observed and expected results seen were due to chance. ✓

MARKER NOTE

Lucie should include how coat colour and texture are inherited, i.e. Mendelian genetics therefore applies and lizard bobbing speed and throat patch colour genes are not linked. Patch colour is controlled by a dominant red allele and a recessive yellow allele, head bobbing speed is controlled by a dominant fast allele and a recessive slow allele.

Lucie achieves 10/12 marks

Ceri's answer

a) Ff Rr x ffrr ✓

FR, Fr, fR, fr x ffrr ✓

	fr
FR	FfRr
Fr	Ffrr
fR	ffRr
fr	ffrr ✓

ratio is 1:1:1:1 ✓

MARKER NOTE

Ceri should include phenotypes either in the table or in the ratios.

b)

Category	Observed (O)	Expected (E)	$O - E$	$(O - E)^2$	
fast bobbing red patch	27	25	2	4	
fast bobbing yellow patch	22	25	−3	9	
slow bobbing red patch	28	25	3	9	
slow bobbing yellow patch	23	25	−2	4	
Σ	100	100			

✓

$\dfrac{26}{100} = 0.26$

$X^2 = 0.26$ ✗

MARKER NOTE

Ceri has summed $(O - E)^2$ and then divided this by the sum of E rather that working out each $(O - E)^2$ / E and summing these. The calculated chi squared value is wrong as a result. This is a common mistake.

c) Because the calculated value 0.26 is less than the critical value of 7.82, any differences between observed and expected results seen were due to chance. ✓

MARKER NOTE

Ceri must include the level of probability used, i.e. p = 0.05, as 7.82 is also the value for 2 degrees of freedom at p = 0.02.

MARKER NOTE

Ceri has been awarded error carried forward – even though the calculated value is wrong, this was penalised in part b) so is not penalised again in part c). Ceri needs to include a null hypothesis, and should include how patch colour and head bobbing speed are inherited, i.e. Mendelian genetics therefore applies and patch colour and head bobbing speed genes are not linked, etc.

Ceri achieves 6/12 marks

EXAM TIP

When performing statistical tests, include a null hypothesis and ensure that results are explained in terms of significance, chance and the probability value 0.05.

4.4 Variation and evolution

Topic summary

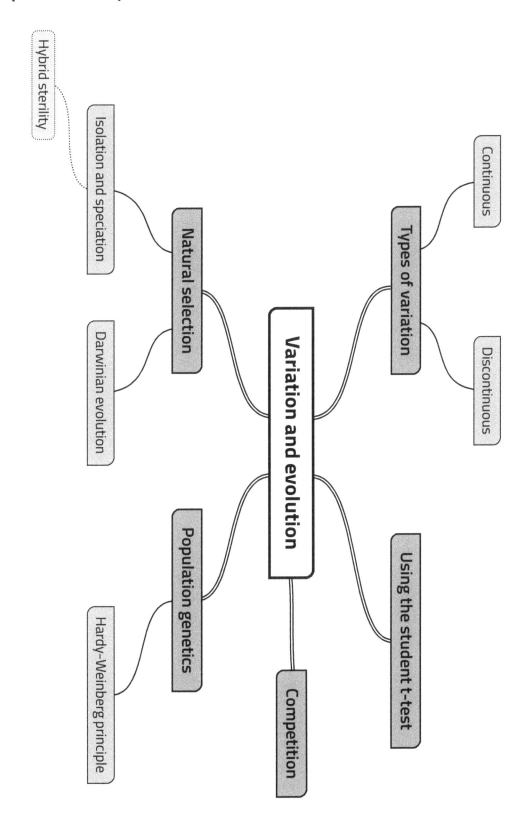

Practice questions

[AO2, AO1]

Q1 ABO blood group distribution is shown in the graph below. Scientists have concluded that blood group is a form of discontinuous variation controlled by monogenic inheritance.

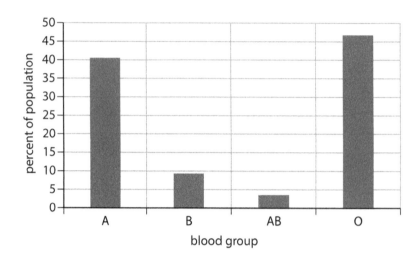

a) Explain what is meant by monogenic. (1)

....

....

b) Using the graph, explain what evidence there is that blood group is a form of discontinuous variation. (1)

....

....

c) Describe *four* ways in which variation can arise. (4)

....

....

....

....

....

....

Q2

[AO1, S]

Distinguish between the following terms:

a) Interspecific and intraspecific competition. (2)

....

....

....

b) Allele frequency and gene pool. (2)

....

....

....

c) Species and speciation. (2)

....

....

....

d) Allopatric and sympatric speciation. (2)

....

....

....

Q3

[M, AO2, AO1]

In response to pathogens, plants have evolved resistance genes whose products convey resistance to a specific virus, bacteria, fungus or insects. These include the ability to produce enzymes, e.g. chitinases, and antimicrobial compounds, e.g. phytoalexins, which are synthesised and accumulate rapidly at areas of pathogen infection. Plants have also evolved to live in arid environments by developing features such as thick waxy cuticles which reduce water loss through transpiration. Over time the frequency of these alleles in populations has changed according to the selection pressures.

On an isolated island, palm plants have genes for a waxy cuticle thickness and disease resistance. Thin waxy cuticle (T) is dominant over thick waxy cuticle (t), and the ability to synthesise phytoalexins (p) is recessive to not being able to synthesis phytoalexins (P). Assume the gene pool is restricted and no new alleles can be introduced to the population by immigration or lost by emigration.

a) Use the Hardy–Weinberg principle to calculate the frequency of heterozygotes carrying the phytoalexin allele in the population of palms, where the number of disease-resistant plants was 1 in 200. Show your working. (5)

The Hardy–Weinberg equation is $p^2 + 2pq + q^2 = 1$

Where:

p^2 = frequency of homozygous dominant alleles

$2pq$ = frequency of heterozygotes

q^2 = frequency of homozygous recessive alleles

Frequency of heterozygotes = ...

b) Using your calculation above, estimate the number of homozygous dominant plants in a population of 10,000. (2)

Answer = ...

c) State *three* assumptions which are made when using the Hardy–Weinberg principle. (3)

....

....

....

d) Suggest how the dominant alleles could be lost from the population over time. (4)

....

....

....

....

....

....

....

Q4

[AO1]

Give an account of the ways in which isolation can lead to speciation. [9 QER]

[AO1, AO2]

Q5 The following illustrations show the evolution of the horse over the past 55 million years, and a horseshoe crab today compared with a 400-million-year-old fossil.

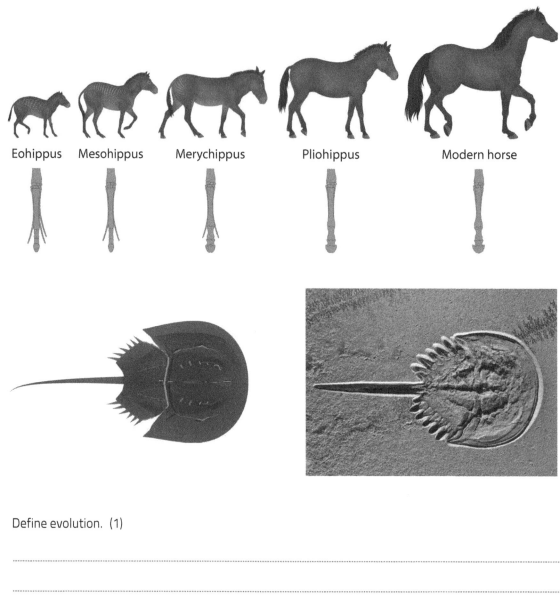

a) Define evolution. (1)

....

....

b) Suggest reasons for the differences seen between the evolution in the horse and the horseshoe crab. (5)

....

....

....

....

....

....

....

[AO2]

Q6 The following photograph shows a grey donkey (chromosome number 62) and black horse (chromosome number 64). These are regarded as two separate species which can interbreed to produce a mule.

a) Using the information provided, explain why these animals are regarded as two separate species. (5)

....

....

....

....

....

....

....

....

b) Suggest the process by which the two species may have arisen. (2)

....

....

....

Question and mock answer analysis

Q&A 1 [AO1, AO2]

Cystic fibrosis is a recessive condition affecting around 1 in 2500 human babies in the UK. The Hardy–Weinberg formula states that if alleles A and a are present in a population with the frequencies of p and q, the proportion of individuals homozygous for the dominant allele (AA) will be p^2, the proportion of heterozygotes (Aa) will be 2pq, and the proportion of homozygous recessives (aa) will be q^2, where p + q = 1.

The Hardy–Weinberg equation is $p^2 + 2pq + q^2 = 1$

a) What is meant by the term recessive allele? (2)

b) Use the Hardy–Weinberg formula to estimate the number of carriers of cystic fibrosis per 1000 in the UK. Show your working. (4)

c) State two conditions which should exist under ideal conditions for the Hardy–Weinberg principle to apply in this example. (2)

Lucie's answer

a) An allele that is only expressed in the homozygous recessive, e.g. aa. ✓

b) $aa = \dfrac{1}{2500} = 0.0004 = q^2$ ✓

$q = \sqrt{0.0004} = 0.02$

$p = 1 - 0.02 = 0.98.$ ✓

$Aa = 2pq = 2 \times 0.02 \times 0.98$ ✓ $= 0.039$ or 39 per 1000 population. ✓

c) Allele frequencies are the same in both sexes ✓ and reproduction is random ✓

MARKER NOTE
Lucie also needs to define allele, i.e. a different form of the same gene (a gene is a section of DNA that codes for a specific polypeptide).

Lucie achieves 7/8 marks

Ceri's answer

a) A recessive allele codes for a protein. ✗

and the phenotype is only seen when both copies of the allele are present, e.g. aa. ✓

MARKER NOTE
Ceri needs to correctly define allele – that it is a different form of the same gene.

b) $q^2 = \dfrac{1}{2500} = 0.0004$ ✓

$q = 0.02$

$p = 0.98.$ ✓

proportion of population that are carriers are 0.98 ✗

MARKER NOTE
Ceri should show workings more fully to enable marks for process to be given. The final step of calculating the value of 2pq (2 x 0.02 x 0.98) to get 0.039 and then multiplying by 1000 to get the proportion of the population per 1000 needs to be included.

c) The population size is very large ✓ and there is no migration. ✗

MARKER NOTE
In this example in humans, migration is extremely likely as people migrate to and from the UK each year, so Ceri should have used another example, e.g. there is no selection, or mating is random.

EXAMINER COMMENTARY
It is important to answer in context of the question, as some examples will not apply. Always show your full working so marks for process can be awarded.

Ceri achieves 4/8 marks

4.5 Application of reproduction and genetics

Topic summary

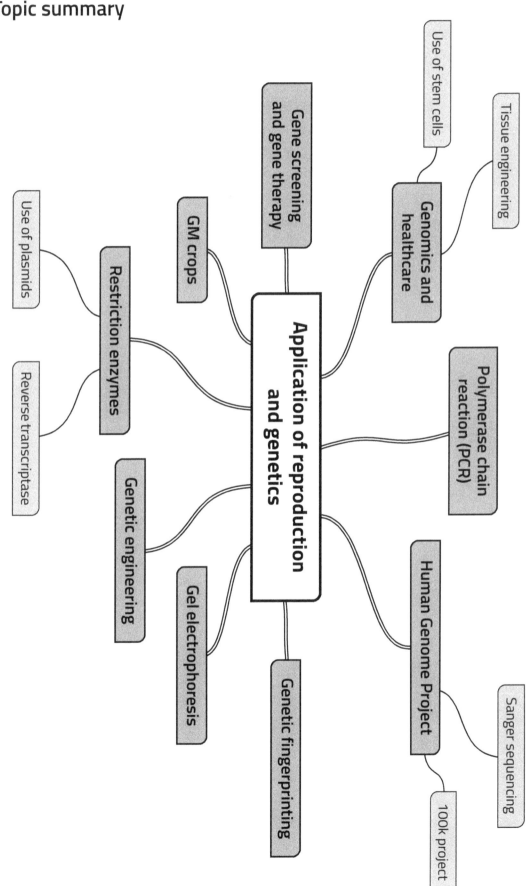

Practice questions

[AO2, AO1]

Q1

D7S280 is a repeating sequence found on human chromosome 7. The DNA sequence of a representative allele of this locus is shown below. The tetrameric repeat sequence of D7S280 is 'gata'. Different alleles of this locus have from 6 to 15 tandem repeats of the 'gata' sequence. The following DNA sequence shows the tetrameric repeat sequence of D7S280:

   1   aatttttgta ttttttttag agacggggtt tcaccatgtt ggtcaggctg actatggagt

  61   tattttaagg ttaatatata taaagggtat gatagaacac ttgtcatagt ttagaacgaa

 121 ctaac**gatag atagatagat agatagatag atagatag agatagatag atagata**gat

 181 tgatagtttt tttttatctc actaaatagt ctatagtaaa catttaatta ccaatatttg

 241 gtgcaattct gtcaatgagg ataaatgtgg aatcgttata attcttaaga atatatattc

 301 cctctgagtt tttgatacct cagattttaa ggcc

a)   An amplified sample of the DNA sequence was run on an electrophoresis gel against a DNA ladder of known size. On the diagram below, mark a band and label it sample 1, to show where you would expect the band of D7S280 to appear. (1)

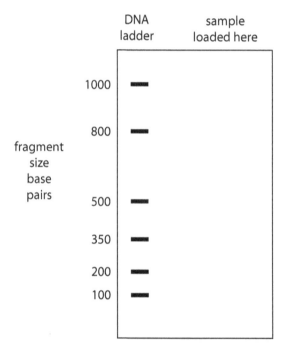

b)   Mark on the diagram where the positive electrode (+) and negative electrode (−) would be found. Explain your choices. (3)

Explanation

....

....

....

c) Combination of short tandem repeats (STR) like the one shown in part a) can be used to produce genetic fingerprints. The diagram below shows the results of a paternity case. Identify the father and explain your choice. (2)

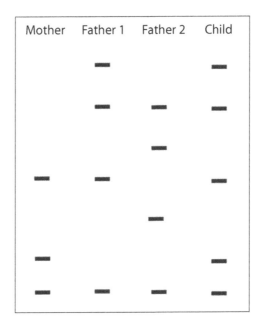

Father = ...

Reason

....

....

d) In order to amplify DNA to have sufficient quantities to run on an agarose gel using electrophoresis, PCR is used.

i) State what is meant by PCR. (1)

....

ii) PCR involves three different temperature steps. State the temperature used, and explain the purpose of each temperature. (5)

Temperature 1 ....

Purpose ....

Temperature 2 ....

Purpose ....

Temperature 3 ....

Purpose ....

[M, AO2]

Q2 The diagram below shows a bacterial plasmid 2105 base pairs long, and the positions where seven different restriction enzymes cut it, e.g. Hind II site is found at 350 base pairs from the Bam I site. Using the diagram, answer the following questions.

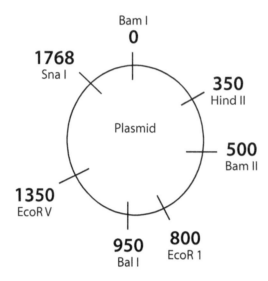

a) Calculate the number of fragments produced if all seven restriction enzymes were used.

Answer ....

b) Calculate the size of the largest fragment. Show your working. (2)

Answer ....

c) The plasmid was cut using Bam I and Sna I. Calculate the size of the fragments produced. Show your working. (2)

Answer

....

d) A gene was found between 1015 and 1750 bases. State which restriction enzymes would need to be used to remove the gene with as few extra bases as possible. (1)

Answer

....

Q3 [S, AO1, AO2]

Sickle cell disease is common in Afro-Caribbean, Middle Eastern, Eastern Mediterranean and Asian populations that evolved in malarial habitats. In sickle cell, a mutation involves a substitution of adenine for thymine which results in the change of one amino acid to valine, causing red cells to deform and block capillaries under low partial pressures of oxygen.

a) Name the type of mutation shown. (1)

b) Name the group of nucleotides to which adenine and thymine belong. (1)

c) The alleles for normal and affected haemoglobin are co-dominant. Explain why this is an example of heterozygote advantage. (2)

d) Using your knowledge of sickle cell disease and gene technology, explain how you could treat a sufferer of sickle cell. (4)

[S, AO2, AO1]

Q4 Sanger sequencing is a method of sequencing DNA, named after the scientist who invented it. It works by sequencing small fragments of DNA around 800 bases in length created by the use of restriction enzymes. DNA polymerase is then used to synthesise complementary strands using the polymerase chain reaction (PCR). Four reactions were carried out (one for adenine, thymine, cytosine and guanine), each containing complementary nucleotides marked with a radioactive marker, but a proportion of the nucleotides used in each reaction had been altered and are called stop nucleotides. When these were incorporated into the complementary strand, further synthesis was prevented. When the results for all the reactions for each nucleotide are run out side by side on an agarose gel using electrophoresis, and the resulting gel exposed to x-ray film to detect the radioactive signal, the sequence can be determined by reading the banding pattern.

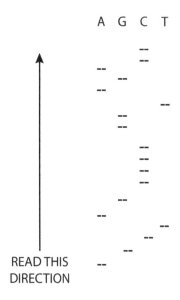

Using the gel results above and your knowledge of gel electrophoresis, answer the following questions.

a) Suggest why the gel is read in the direction shown. (2)

....

....

....

b) Work out the DNA sequence of the sample. (2)

....

c) Describe **three** limitations of PCR. (3)

....

....

....

....

d)   PCR involves three steps. Step 2 requires cooling to 50–60 °C to allow primers to attach to the template by complementary base pairing. Suggest why scientists may need to use a lower temperature for some primers. (3)

....

....

....

....

e)   Describe *three* ethical concerns with the Human Genome Project. (3)

....

....

....

....

Unit 4 Variation, Inheritance and Options

Q5

[AO1]

Outline the pros, cons and hazards of genetically engineering bacteria. [9 QER]

Q6

[AO2, AO1]

The bacterial plasmid shown below contains two marker genes: the first one is for ampicillin resistance, the second contains the Lac Z gene which metabolises the chemical x-gal, turning it from colourless to blue. Colonies grown on a plate spread with x-gal appear blue rather than white.

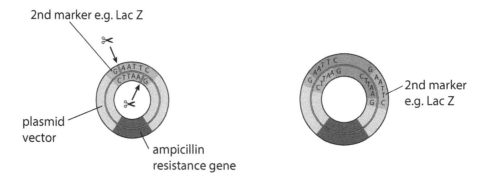

a) Explain the function of both marker genes. (3)

....

....

....

....

....

b) Describe the steps needed to insert DNA at the cloning site. (4)

....

....

....

....

....

Question and mock answer analysis

Q&A 1

[a = AO1, b & c = AO3]

Scientists started mapping a section of DNA by digesting it with different restriction enzymes and estimating the size of each fragment by running the products out on an agarose gel alongside a DNA ladder containing DNA fragments of known sizes. The results are shown in the table below:

Enzymes used	Estimated sizes of fragments produced / base pairs
EcoRI	550, 450
BamHI	750, 300
SnaI	500, 325, 200
EcoRI and PstII	550, 450
EcoRI and HindIII	550, 250, 200

a) What is a restriction enzyme? (1)

b) Using a DNA ladder to estimate size of DNA fragments has its limitations and is often inaccurate. What evidence is there in the data to support this claim? (2)

c) Draw conclusions from the results, justifying your answer. (3)

Lucie's answer

a) A bacterial enzyme that cuts single-stranded DNA at a specific base pair sequence. ✓

b) The same DNA was cut with different enzymes and the total size of the fragments produced by each reaction was different ✓, e.g. ECORI produced fragments totalling 1000, whereas BamHI totalled 1050 even though the DNA used was the same. ✓

c) The size of the DNA fragment is 1000 because the fragments produced in all the digests total approximately 1000. ✓ PstII does not cut the DNA so there is no recognition sequence for PstII in the sample, because the number and size of fragments produced are the same as when using ECORI alone. ✓

HindII cuts within the 450 bp ECORI fragment because when both enzymes are used the 450 bp fragment is no longer present, but two fragments totalling 450 bp are present. ✓

MARKER NOTE

Lucie could also conclude that EcoRI and BamHI only cut the DNA once, as two fragments are produced. SnaI cuts twice as three fragments are produced.

Lucie achieves 6/6 marks

Ceri's answer

a) An enzyme that cuts DNA.

MARKER NOTE
The definition needs further detail, e.g. Cuts DNA at a specific recognition sequence.

b) Different sized fragments are produced when different enzymes are used.

MARKER NOTE
Ceri needs to include specific examples to support the answer, e.g. fragments total 1000 base pairs when cut with EcoRI but 1025 when cut with Snal.

c) The DNA must be approximately 1000 bases long.

MARKER NOTE
The conclusions are valid but should include justification, e.g. DNA is approx. 1000 bases long because the fragments produced add to 1000 with EcoRI but 1025 with Snal.

There is no site for PstII to cut within the DNA, because the result is the same when using EcoRI and EcoRI & PstII. ✓ The other enzymes cut once but Snal must cut twice because three fragments are produced. ✓

Ceri achieves 2/6 marks

EXAM TIP
Always use evidence to justify conclusions.

Option A: Immunology and disease

Topic summary

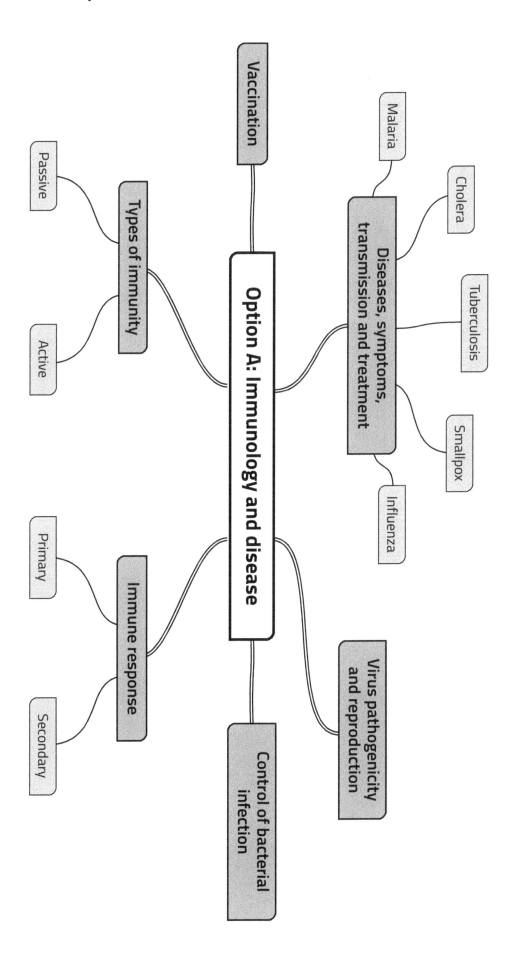

Practice questions

[AO1]

Q1 Distinguish between the following terms.

a) Epidemic and endemic. (2)

b) Antigen and antibody. (2)

c) Bactericidal and bacteriostatic, giving examples. (3)

d) Passive and active immunity, explaining the advantages and disadvantages of each. (4)

Q2

[AO2, AO1, S]

Rifampicin is a broad-spectrum antibiotic which specifically inhibits bacterial DNA-dependent RNA polymerase, the enzyme responsible for DNA transcription. It is bactericidal to both intracellular and extracellular Gram-positive and Gram-negative bacteria and has activity against Mycobacterium tuberculosis. It is often used in combination with other antibiotics to treat tuberculosis, which enables the combination therapy to be used over a shorter six-month course.

a) Suggest how Rifampicin kills bacteria. (3)

b) Distinguish between the mode of action of rifampicin and tetracycline. (3)

c) Suggest why a combination therapy is used to treat tuberculosis. (2)

[AO1, AO2]

Q3 There are five classes of antibodies (immunoglobulins) which are produced in response to different antigens. Two of the most common produced in response to a virus are IgM and IgG. IgM is found mainly in the blood and lymph fluid, and is the first antibody to be made by the body to fight a new infection, whereas IgG is the most abundant type of antibody. It is found in all body fluids and provides protection against both viral and bacterial infections. The graph below shows levels of IgM and IgG following a viral infection.

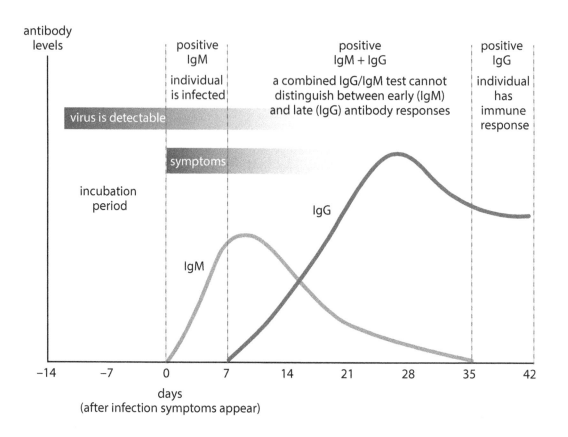

a) Name the type of response shown. Explain your answer. (2)

....

....

....

b) In the space below, draw a fully annotated diagram of an antibody molecule. (3)

c) Describe the processes which occur during the response you named in part a). (4)

....

....

....

....

....

....

d) Using the graph, the information provided, and your knowledge, suggest the role of IgM and IgG in the immune response. (5)

....

....

....

....

....

....

....

....

Unit 4 Variation, Inheritance and Options

Question and mock answer analysis

[a & b = AO1, c = AO2]

The graph shows the antibody concentration in the blood following two exposures to the same antigen:

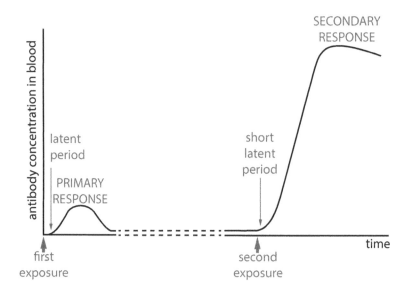

a) State how antibodies are produced. (1)

b) Explain why the antibody concentration in the blood is higher following the second exposure than the first exposure. (3)

c) All antibodies have the same Y-shaped molecule. Explain how they are able to bind to specific antigens. (2)

Lucie's answer

a) Antibodies are made by the humoral response by the B-lymphocytes ✓

b) Following first exposure, macrophages need to engulf the foreign antigen and incorporate the antigens into their own cell membranes during the time called the latent period. ✓

> **MARKER NOTE**
> Lucie should include that T helper cells secrete cytokines which trigger B plasma cells to produce antibodies, which takes time.

Following second exposure, memory cells undergo clonal expansion much faster than following the first exposure because antigen presentation doesn't occur, so antibodies are made much more quickly and in larger quantities. ✓

c) The antigen binding site contains a variable polypeptide chain ✓ which has a specific shape complementary to the antigen. ✓

Lucie achieves 5/6 marks

Ceri's answer

a) *Antibodies are made by the humoral response*

MARKER NOTE
Ceri should provide a fuller answer to include the B-lymphocytes.

b) *Concentration is higher following second exposure because memory cells undergo clonal expansion much faster than following the first exposure because macrophages don't need to engulf the foreign antigen and incorporate the antigens into their own cell membranes.* ✓

MARKER NOTE
Ceri needs to include that during the primary response T helper cells need to secrete cytokines to trigger B plasma cells to produce antibodies which takes time. Because the secondary response occurs more quickly, a higher level of antibody production can be achieved.

c) *Each antibody is different and has a complementary shape to the antigen.* ✓

MARKER NOTE
Ceri fails to explain how this is achieved, i.e. due to a variable polypeptide region at the antigen binding site.

Ceri achieves 2/6 marks

EXAM TIP
It is important to make a comparison when differences are asked for, and to include examples.

Unit 4 Variation, Inheritance and Options

Option B: Human musculoskeletal anatomy

Topic summary

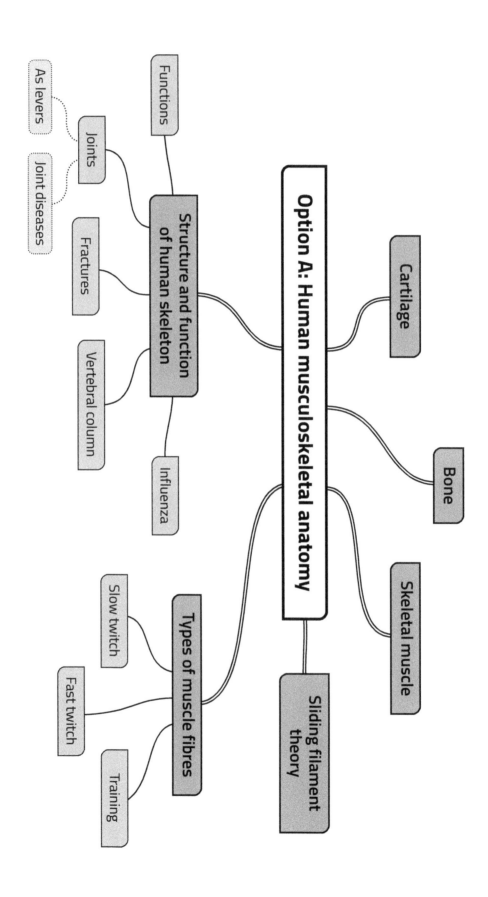

Practice questions

Q1

[AO2, AO1, S]

The following diagrams show the structure of yellow elastic cartilage and fibrocartilage:

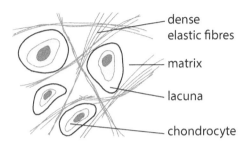

Yellow elastic cartilage

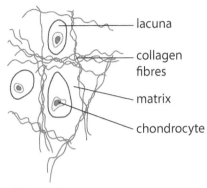

Fibrocartilage

a)   State the tissue type to which cartilage belongs. (1)

....

b)   Explain how the structures of each shown in the diagram enable it to perform its function. (4)

....

....

....

....

....

c)   Explain why damaged cartilage takes a long time to heal. (2)

....

....

....

[AO2, AO1]

Q2 The image below represents a portion of a myofibril.

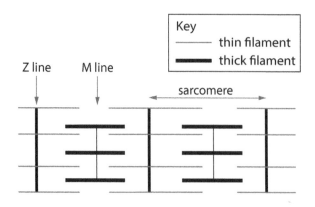

a) Label on the diagram the A and I bands, and H zone. (2)

b) State what happens to the A and I bands and the H zone you have labelled, during muscle contraction. (2)

c) Explain the role of ATP in muscle contraction. (3)

d) Calcium channel blockers are used to treat cardiovascular conditions such as hypertension, by causing arteries to dilate, reducing pressure within and making it easier for the heart to pump blood. Using your knowledge of the sliding filament theory, suggest their mode of action. (4)

[M, AO1]

Q3 A biceps curl is an example of a 3rd order level. The diagram below shows the forces involved:

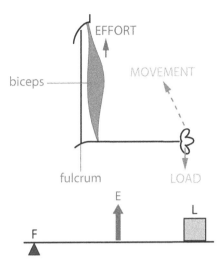

When a lever is at equilibrium, $F_1 \times d_1 = F_2 \times d_2$

Where F^1 = load, d_1 = distance from the fulcrum (elbow) to the load, F_2 = effort, and d_2 = distance from fulcrum (elbow) to the insertion of the biceps muscle. Assume 1 kg = 9.8 newtons.

a) State what is meant by a lever. (1)

....

b) Calculate the force exerted by the effort needed to hold the load level if load = 20 kg, distance from the fulcrum (elbow) to the load is 0.38 m, distance from fulcrum (elbow) to the insertion of the biceps muscle is 0.05 m. Show your working. (2)

Answer ....

c) Calculate the maximum load which could be held if the maximum lifting strength of the muscle (the effort) is 2500 newtons. Show your working. (2)

Answer ....

Q4

[AO1]

Skeletal muscle is made up of muscle fibres which are long thin cells containing many nuclei. Each fibre contains many myofibrils.

a) Describe two differences between slow twitch and fast twitch muscle fibres. (2)

....

....

....

....

b) The type of training used by marathon runners has been shown to increase the relative proportions of slow twitch fibres. State one other change that occurs in muscles during endurance training and explain the benefit to a marathon runner. (2)

....

....

....

....

c) During exercise the main energy source is muscle glycogen which is stored in muscles. Describe how the body utilises another store once glycogen has been exhausted and before anaerobic respiration begins. (2)

....

....

....

Question and mock answer analysis

[a = AO2, b = AO1]

Q&A 1 The diagram below shows three different sections through a myofibril showing the arrangement of myofilaments:

A

B

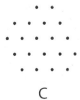

C

a) State which diagram A, B or C represents the H zone. Explain your answer. (2)

b) During strenuous exercise, muscles may temporarily respire anaerobically. Explain why it is important for the muscles of an athlete to convert pyruvate into lactate (lactic acid) and the consequence of a build-up of lactate on muscle contraction. (3)

Lucie's answer

a) Diagram C ✓. This is because only thicker myosin filaments are visible, which are found in the H zone ✓.

b) It allows glycolysis to continue, because NAD is regenerated ✓ when pyruvate is reduced to lactate. ✓ A build-up of lactate would inhibit chloride ions, which regulate muscle contraction, resulting in a sustained contraction leading to cramp. ✓

Lucie achieves 5/5 marks

Ceri's answer

a) Diagram C ✓. This is because it appears darker like the H zone. ✗

> **MARKER NOTE**
> Whilst it may appear darker, Ceri needs to link this to the type of fibres present, i.e. only myosin.

b) It allows for the build-up of an oxygen debt.

> **MARKER NOTE**
> Whilst a build-up of an oxygen debt happens, Ceri needs to go further to explain the consequence, i.e. that NAD is regenerated allowing glycolysis to continue. Ceri also needs to include the consequence of a build-up of lactate on muscle contraction.

Ceri achieves 1/5 marks

EXAM TIP
Read the question carefully and explain your answer fully.

Option C: Neurology and behaviour

Topic summary

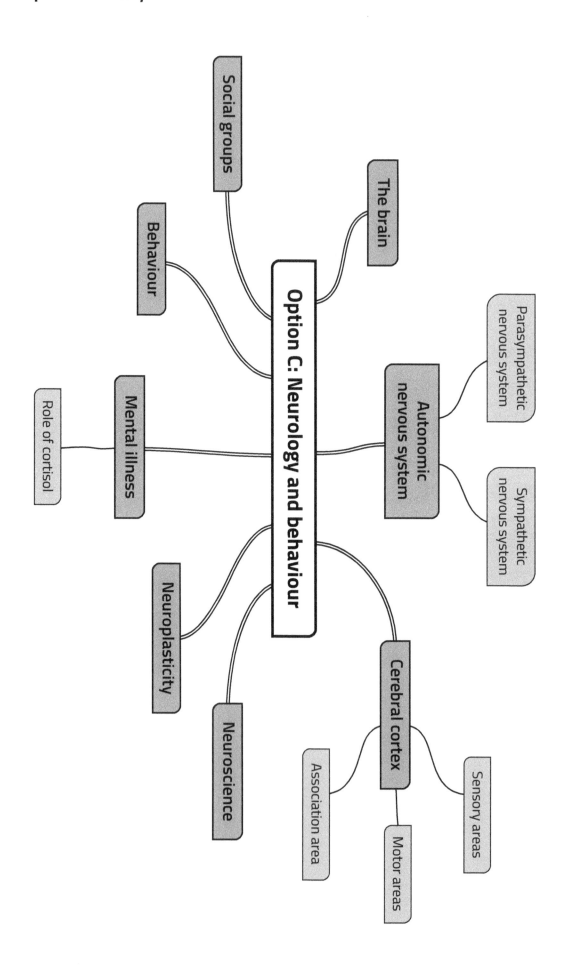

Practice questions

[AO2, AO1, AO3]

Q1 The diagram below shows the functional areas of the brain:

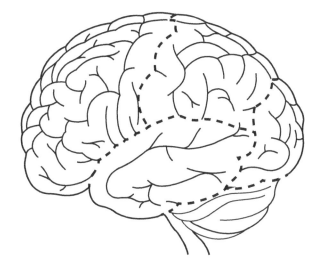

a) Functional magnetic resonance imaging is a technique for examining activity of brain tissue in real time. On the diagram, shade and label the main area(s) with the corresponding letter which you would expect to show a positive signal, indicating activity, following stimulation from:

i)   A – viewing a photograph. (1)

ii) B – tasting food. (1)

iii) C – talking. (1)

b) Distinguish between the following terms:

i)   Sympathetic and parasympathetic nervous systems. (4)

ii) Cerebrum and cerebellum. (2)

c) A student carried out an experiment to measure the effects of stress on cortisol and blood glucose levels in a laboratory rat. The results are shown below:

Action	Cortisol blood levels ng/ml	Blood glucose levels mg/dl
At rest	4	110
30 mins following stress event	16	180
60 mins following stress event	8	305
120 mins following stress event	5	120

i) What conclusions can be reached from the experiment as to the effects of stress on cortisol and blood glucose levels in rats. Explain your answer. (5)

ii) Comment on the reliability of the experiment, suggesting improvements. (3)

[AO1]

Q2

Explain how the different types of behaviour are important to organisms in protecting themselves, finding food, reproduction and developing skills. [9 QER]

[AO1]

Question and mock answer analysis

[a & c = AO1, b = AO2]

Q&A 1

a) Explain the difference between taxes and kineses. (1)

b) A victim of a car crash suffered damage to their forebrain. Explain why this person would have difficulties in forming permanent memories but would be less likely to suffer from stress. (2)

c) Use examples to distinguish between classical and operant conditioning. (2)

Lucie's answer

a) Kineses are non-directional whereas in taxes the direction of the movement is related to the direction of the stimulus either towards or away from it. ✓

b) The hippocampus is located in the forebrain and is involved in consolidating memories into a permanent store. If this were damaged, the person would be unable to do this. ✓ It is also responsible for producing cortisol, the stress hormone.

> **MARKER NOTE**
> Lucie needs to make a clear link between damage to the hippocampus and its role in controlling the production of cortisol from the adrenal glands.

c) Classical conditioning involves the association between a natural and an artificial stimulus to bring about the same response, e.g. a dog associating the ringing of a bell with food, ✓ whereas operant conditioning involves the association between a particular behaviour and a reward or punishment, e.g. mice learning to press a lever to receive food (reward) or to stop a loud noise (punishment). ✓

Lucie achieves 4/5 marks

Ceri's answer

a) Kineses are non-directional; taxes are directional.

> **MARKER NOTE**
> Taxes involve movement that is related to the direction of the stimulus.

b) The hippocampus may have been damaged, which is responsible for forming memories. ✓ It is also responsible for producing cortisol.

> **MARKER NOTE**
> The role of cortisol needs to be included.

c) In classical conditioning, a dog learns to associate the ringing of a bell with food, whereas operant conditioning involves mice learning to press a lever to receive food. ✓

> **MARKER NOTE**
> Ceri should include the main difference between the two types of conditioning, i.e. that classical conditioning involves the association between a natural and an artificial stimulus to bring about the same response whereas operant conditioning involves the association between a particular behaviour and a reward or punishment.

Ceri achieves 2/5 marks

> **EXAM TIP**
> When distinguishing between two terms make sure you include detail on both terms.

Practice papers

Unit 3 Practice paper – Energy, Homeostasis and the Environment

90 Marks, 2 hours

 Q1 ATP belongs to a group of molecules called nucleotides.

a) In the space below, draw a fully labelled diagram of a molecule of ATP. (3)

b) State what is meant by the phosphorylation of glucose, and explain why this makes it easier to split the glucose molecule. (3)

c) Compare the synthesis of ATP in mitochondria and chloroplasts. (4)

Q2 The electronmicrograph below shows a chloroplast:

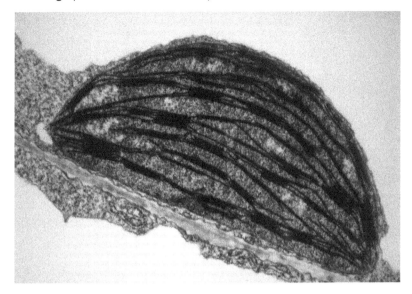

a) On the diagram, label where the light-dependent and light-independent reactions take place. (2)

b) Describe *two* adaptations of a chloroplast to maximise absorption of light. (2)

....

....

....

....

c) Distinguish between cyclic and non-cyclic photophosphorylation. (3)

....

....

....

....

....

....

Q3 The diagram below shows a kidney nephron:

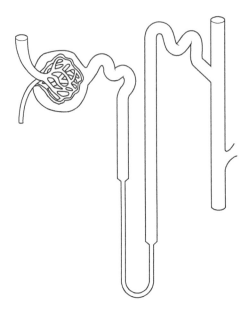

a) On the diagram, label the following structures. (2)

   i) Glomerulus

   ii) Proximal convoluted tubule

   iii) Loop of Henle.

b) Explain three adaptations found in the glomerulus and Bowman's capsule that enable ultrafiltration to occur. (3)

....

....

....

....

....

....

c)   An experiment was carried out to measure the concentration of sodium ions at different places along the nephron. The results are shown below:

Part of nephron	Concentration of sodium ions in the filtrate / mmol dm^{-3}	Percentage of sodium ions remaining / %
Start of proximal convoluted tubule	155	100
End of convoluted tubule	158	55
Start of distal convoluted tubule	55	15

i)   Explain the reasons for the decrease in percentage of sodium ions over the length of the nephron. (3)

....

....

....

....

....

....

ii)   Explain why there is no decrease seen in the concentration of sodium ions through the proximal convoluted tubule. (2)

....

....

....

iii)   Explain the effect an increase in anti-diuretic hormone (ADH) would have on the sodium ion concentration in the blood. (3)

....

....

....

....

....

Q4 An experiment was carried out to investigate the effect of ADH on urine production in a laboratory rat. The rate of urine production was measured every five minutes over a 45-minute period and the results are shown below. Ten minutes after the experiment began, ADH was injected into the vein of the rat.

Time / minutes	Rate of urine production / $mm^{-3} min^{-1}$
0	4.5
5	4.5
10	4.4
15	3.0
20	2.1
25	1.4
30	0.9
35	1.5
40	2.4
45	3.5

a) Suggest why the ADH injection was not administered until ten minutes after the experiment began. (1)

....

....

b) Draw a graph to show the results obtained. (5)

c) Draw conclusions from the results seen. Explain your answer. (5)

....

....

....

....

....

....

....

d) Suggest improvements to the experiment. (2)

....

....

....

....

....

....

....

Q5 A fermenter was used to grow bacteria in a medium containing glucose over a 24-hour period. 1 cm^3 was removed every four hours to calculate the number of bacterial cells and measure the glucose concentration. The results are shown in the graph below:

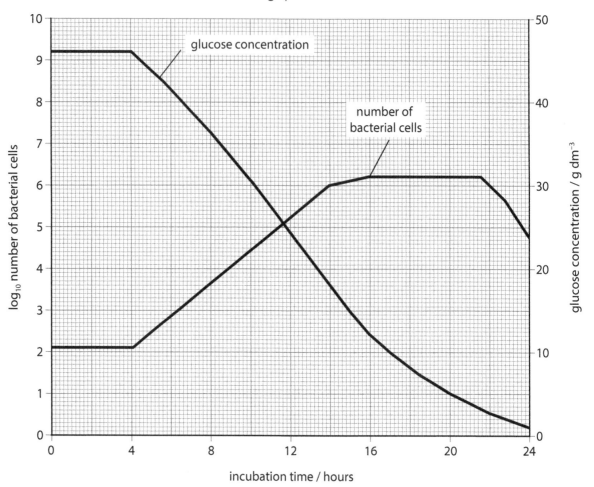

a) Describe the relationship between the numbers of bacterial cells in the culture and the glucose concentration in the medium from 0 to 20 hours. (3)

....

....

....

....

b) Suggest a reason for the relationship seen between bacterial cell number and glucose concentration from 16 to 20 hours. (2)

....

....

....

c)   Calculate the number of generations of bacteria produced between 4 hours and 12 hours, using the formula below. Show your working. (3)

$$n = \frac{\log_{10} N_1 - \log_{10} N_0}{\log_{10} 2}$$

where n = number of generations

N_0 = number of cells at 4 hours

N_1 = number of cells at 12 hours

$\log_{10} 2$ (doubling time) = 0.6

Number of generations = ...

d)   Explain the results seen after 22 hours. (3)

....

....

....

....

....

Q6 The following diagrams show the order of some parts of the oxidative phosphorylation of NADH and FADH in the electron transport chain, and the Krebs cycle.

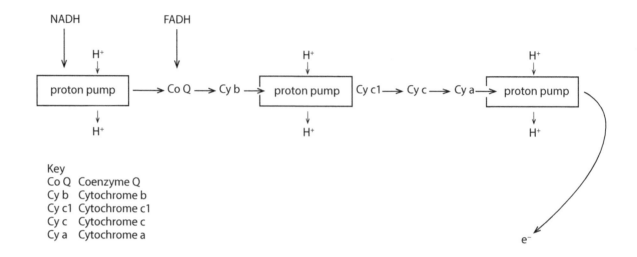

Key
Co Q   Coenzyme Q
Cy b   Cytochrome b
Cy c1 Cytochrome c1
Cy c   Cytochrome c
Cy a   Cytochrome a

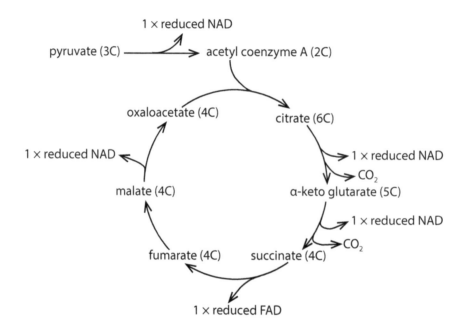

a) State **precisely** where the electron transport chain and Krebs cycle take place. (2)

....

....

b) Using your knowledge and the diagram, explain the difference in ATP yields from NADH and FADH. (6)

....

....

....

....

....

....

....

c) Using your knowledge and ***both*** diagrams, explain why the oxidation of one molecule of succinate only yields 5 ATP. (3)

....

....

....

....

....

d) A high concentration of oxaloacetate inhibits the enzyme which catalyses the conversion of succinate into fumarate. Explain the effects of a high concentration of oxaloacetate on the Krebs cycle and suggest why this might be beneficial to the organism. (6).

....

....

....

....

....

....

....

....

....

....

Unit 3 Practice Paper

Q7 The graph below shows the total greenhouse and carbon dioxide UK emissions from 1990 to 2018:

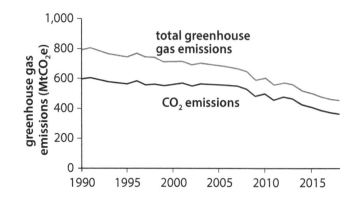

Suggest what measures have been used in the UK to reduce emissions, and how planetary boundaries have been affected as a result. [9QER]

....

....

....

....

....

....

....

....

....

....

....

....

....

....

....

....

....

....

Q8

a) Streak plating is a technique used to transfer large numbers of bacterial cultures to an agar nutrient plate, reducing their numbers with each streak. Describe the technique of streak plating, including the precautions that need to be taken to ensure this is carried out aseptically. (4)

b) 1 cm^3 of a suspension of Gram-positive bacteria was evenly spread onto an agar plate using a glass spreader rod. After 30 minutes, paper discs impregnated with four different antibiotics, I, II, III and IV, were placed carefully onto the plate, ensuring there was an even space between them. The procedure was then repeated using a suspension of a Gram-negative bacteria, and both plates incubated at 30 °C for 48 hours. The appearance of both plates following incubation is shown below:

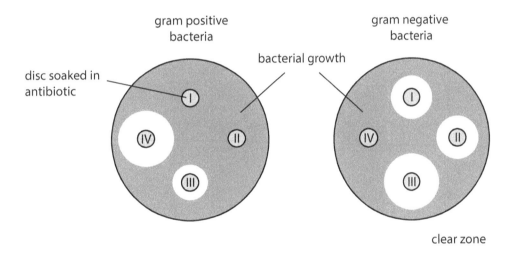

i) How effective was each antibiotic against the two different bacteria used? Explain your answer. (3)

ii) Suggest which antibiotic is penicillin. Give a reason for your choice. (3)

Unit 4 Practice paper – Variation, Inheritance, and Options

Section A 70 marks

Section B Options – choose one section to answer 20 marks

Total 90 marks, 2 hours.

You are advised to spend 25 minutes on Section B

Q1 The diagram below shows the male human reproductive organs:

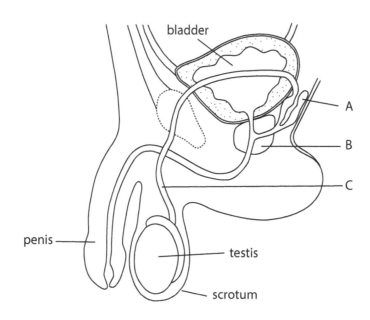

a) Identify the following structures, and state their function. (6)

i) A ....

Function

....

....

ii) B ....

Function

....

....

iii) C ....

Function

....

....

b)   A recent study found the number of men with increased sperm counts was decreasing, and the number of men with low sperm counts was increasing above the average reduced sperm count. Suggest an explanation for the findings, and a consequence of it. (2)

....

....

....

....

c)   The following diagram shows the steps in spermatogenesis.

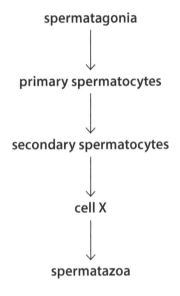

spermatagonia

↓

primary spermatocytes

↓

secondary spermatocytes

↓

cell X

↓

spermatazoa

i)   Name cell X. (1)

....

ii)   Mark clearly on the diagram which cell(s) are haploid. (1)

iii)   Explain how the process of cell division between spermatogonia and primary spermatocytes differs from that between primary and secondary spermatocytes. (3)

....

....

....

....

iv)   Explain how spermatozoa are adapted for their function. (3)

....

....

....

Q2 Phenylketonuria (PKU) is a rare inherited disorder, where sufferers are unable to break down the amino acid phenylalanine, which then builds up in their blood and brain, which can lead to brain damage. At around 5 days old, babies are offered new-born blood spot screening to test for PKU, which involves pricking the baby's heel to collect drops of blood to test.

a) In the space below, draw the structure of a typical amino acid molecule. Label the functional groups. (3)

b) PKU arises from a point mutation in the PAH gene. To date, over 520 different mutations have been identified.

  i) Explain what is meant by the term point mutation and its consequence. (4)

  ii) Explain why a point mutation may occur in a person's DNA, but not necessarily lead to a disease like PKU. (3)

c) The family tree below shows the incidence of PKU in a family:

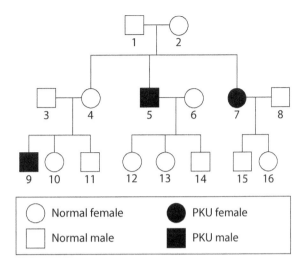

i) What can you conclude about the inheritance of PKU? Explain your answer. (3)

....

....

....

....

ii) Calculate the chance of parents 1 and 2 having a male child with PKU. (2). Explain your answer.

....

....

....

d) Explain how PCR could be used to test for the presence of PKU. (4)

....

....

....

....

....

....

Q3 In the Pleistocene age (2.8 million years to 11,700 years ago) the closure of the Isthmus of Panama separated the basins of the Eastern Pacific and the Caribbean Sea, which resulted in the evolution of two different species of butterflyfishes (Chaetodontidae), the scrawled butterflyfish (*Chaetodon meyeri*) and the ornate butterflyfish (*Chaetodon ornatissimus*).

Scrawled butterflyfish

Ornate butterflyfish

The scrawled butterflyfish is distributed primarily in the Indian Ocean whilst the ornate butterflyfish is distributed primarily in the Central-West Pacific. It has been found that both species are descended from a common ancestor before the isthmus formed.

a) Using the information provided, explain why *Chaetodon meyeri* and *Chaetodon ornatissimus* can be regarded as closely related species. (3)

....

....

....

....

b)   Outline two further tests which could be performed to confirm they are two different species. (2)

....

....

....

....

c)   Using the information provided, explain how the two different species could have arisen. (5)

....

....

....

....

....

....

....

....

Q4 Fruit flies (*Drosophila melanogaster*) can be found with either a brown or black coloured body, and with either long or vestigial wings. The allele for long wings is the dominant phenotype, and short, misshapen wings called vestigial wings, are recessive. Brown body colour is dominant over black.

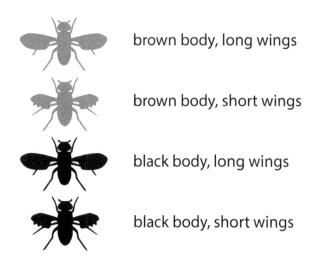

brown body, long wings

brown body, short wings

black body, long wings

black body, short wings

a) Complete the genetic diagram below to show the genotypes and phenotypes expected from a cross between a heterozygous fruit fly with brown body colour and long wings crossed with a homozygous fruit fly with black body colour and vestigial wings. (5)

Parental genotype. .... X ....

Gametes .... X ....

F1 genotypes

F2 phenotypes

Phenotype ratio

b) When a heterozygous fruit fly with brown body colour and long wings was crossed with a homozygous fruit fly with black body colour and vestigial wings, the first generation had 26 brown-bodied flies with long wings, 6 brown-bodied flies with vestigial wings, 5 black-bodied flies with long wings to 23 black-bodied flies with vestigial wings. Using the table below calculate χ^2 for the results of the cross. (3)

Category	Observed (O)	Expected (E)			

Using the formula $\chi^2 = \dfrac{\Sigma\,(O-E)^2}{E}$

$\chi^2 =$ ....

c) Use your calculated χ^2 value and the probability table to conclude how body colour and wing size are inherited. (4)

Degrees of freedom	p = 0.10	p = 0.05	p = 0.02
1	2.71	3.84	5.41
2	4.61	5.99	7.82
3	6.25	7.82	9.84
4	7.78	9.49	11.67
5	9.24	11.07	13.39

....

....

....

....

....

....

Q5 Paclobutrazol (PBZ) is a plant growth retardant which inhibits gibberellin biosynthesis, reducing internodal growth to give stouter stems, by reducing the risk of lodging (stems falling over) in cereal crops, and by increasing the number and weight of fruits per tree. Its mode of action is shown below. PBZ works by inhibiting the oxidation of *ent*-kaurene to *ent*-kauronoic acid by inactivating cytochrome P450-dependent oxygenase.

geranyl diphosphate

↓

ent-kaurene

↓

ent-kauronic acid

↓

GA12 aldehyde

↓

gibberellins

PBZ

An experiment was carried out to investigate the effect of PBZ on germination and growth in cress seeds. Twelve cress seeds were placed on a piece of filter paper in a Petri dish, and watered with distilled water, 10, 50 and 90 mg dm^{-3} PBZ, and left to germinate for ten days. Percentage germination and growth of seedlings were measured every 24 hours, and the seedlings were re-watered. The results are shown below:

Concentration of PBZ / mg dm^{-3}	Number of cress seedlings germinated	Percentage germination / %	Height of seedlings / cm	Average height of seedlings / cm
0	10		8.1, 7.9, 8.0, 7.5, 8.4, 7.5, 8.2, 6.1, 5.9, 8.9	
10	10		5.6, 5.5, 6.4, 6.0, 8.0, 5.3, 5.2, 5.7, 4.9, 5.0	
50	5		3.3, 6.9, 3.4, 2.9, 3.1, 0.0, 0.0, 0.0, 0.0, 0.0	
90	0		0.0, 0.0, 0.0, 0.0, 0.0, 0.0, 0.0, 0.0, 0.0, 0.0	

a) Complete the table to show percentage germination, and average height of seedlings to one decimal place. (2)

b) Identify the independent variable and dependent variables in this experiment. (2)

Independent variable

....

Dependent variable

....

c) Using your knowledge of germination and the information provided, draw conclusions from the experiment, comment on the accuracy of the results, and suggest improvements. (QER 9)

Section B Option papers

Answer **one** section only

20 Marks

You are advised to spend 25 minutes on this section

Option A Immunology and disease

Q6 Acquired immunodeficiency syndrome (AIDS) is caused by the Human Immunodeficiency Virus (HIV). The HIV virion enters macrophages and T helper cells and is able to replicate inside, paralysing one of the main components of the immune system, and is able to hide from anti-HIV antibodies. Upon entry into the target cell, the viral RNA genome is converted into double-stranded DNA. The resulting viral DNA is then imported into the cell nucleus and integrated into the cellular DNA by a virally encoded enzyme, integrase. In the advanced stages of HIV infection, loss of functional T helper cells leads to the symptomatic stage of infection known as AIDS.

Electron micrograph of HIV particle

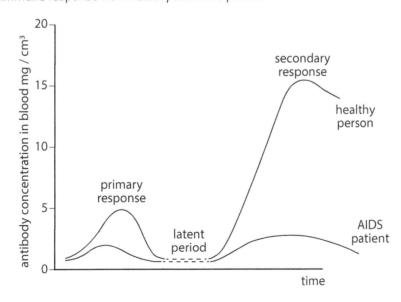

Immune response from healthy and AIDS patient

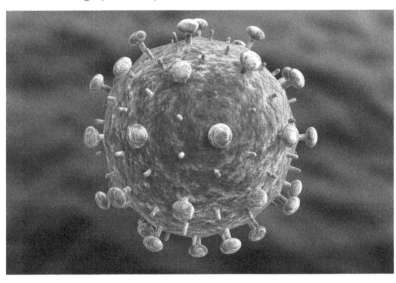

a) Name the type of response which produces antibodies. (1)

....

b) State the total number and type of polypeptide chains which make up the structure of an antibody. (2)

....

....

c) Calculate the percentage increase in antibody levels between the peak of the primary and secondary responses. Show your working. (2)

Answer ....

d) Explain the role of macrophages in the immune response. (2)

....

....

....

e) Using your knowledge of the immune response and the information provided, answer the following questions.

i) Explain why in later stage AIDS, the body is unable to fight off common infections. (4)

....

....

....

....

....

....

ii)   Explain why vaccinating AIDS patients with the influenza vaccine would not be advisable. (4)

f)   Nucleoside reverse transcriptase inhibitors (NRTI) are a class of antiretroviral drugs which are nucleotide base analogues. They function as chain-terminators during the extension of the DNA chain during the reverse transcription process. The NRTI compounds permit correct base-pairing and incorporation into the DNA chain; however, an important hydroxyl group required for addition of the next nucleotide has been replaced by a non-reactive chemical group.

i)   Explain how NRTI drugs prevent HIV replication. (3)

ii)   Explain why penicillin would be ineffective against HIV. (2)

Option B Human musculoskeletal anatomy

Q7 Osteoarthritis is a condition that causes joints to become painful and stiff. It is the most common type of arthritis in the UK. The main symptoms of osteoarthritis are joint pain and stiffness, and problems moving the joint. Some people also have symptoms such as swelling and tenderness. The picture below is an MRI scan showing the cartilage degeneration in a patient suffering from osteoarthritis.

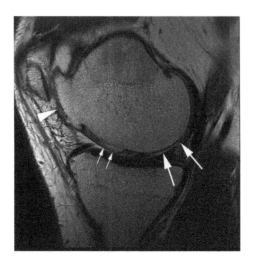

a)   What type of tissue is cartilage? (1)

....

b)   Name the type of cartilage found on articular surfaces. (1)

....

c)   State the type of joint found in the knee. (1)

....

d)   Name the type of cartilage found in intervertebral discs, and explain how its structure and function differs from cartilage found on articular surfaces. (3)

....

....

....

e) To improve joint pain from osteoarthritis, doctors recommend low impact exercise to strengthen muscles around the joint. The diagram below shows the forces at work when performing a squat from a seated position:

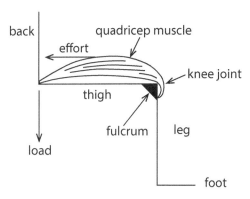

Use the formula below to calculate the force needed to perform a squat (F_2), where F_1 is the load, 70 kg, the distance from the fulcrum to the load (d_1) is 40 cm, the distance from fulcrum to the insertion of the quadriceps muscle (d_2) is 3 cm. Show your working. (2)

$F_1 \times d_1 = F_2 \times d_2$

1 kg = 9.8 newtons, N.

Answer ....

f) Fast twitch muscle fibres rely on anaerobic processes and produce lactic acid, so they tire more quickly than slow twitch fibres. There is a second type of fast twitch muscle fibre called type x, which are used for flight or fight responses, allowing animals to rapidly escape danger. These are even faster and more powerful than type a, but they are also more inefficient, fatiguing very quickly. The table below shows key characteristics of the three types of muscle fibres:

Characteristics	Slow twitch fibres	Fast twitch a fibres	Fast twitch x fibres
Myosin ATPase	Low	High	Highest
Lactate removal rate	Low	Highest	High
Blood capillaries per fibre	High	Moderate	Low
Fibres per motor unit	<300	>300	>300
Creatine phosphate	Low	High	Highest

Using your knowledge and the table above, answer the following questions.

i) Suggest why fast twitch x fibres have a low number of blood capillaries and the highest levels of creatine phosphate and myosin ATPase. (4)

....

....

....

....

ii) Suggest why slow twitch fibres have the lowest levels of creatine phosphate and fibres per motor unit. (3)

....

....

....

....

g) Explain the effects and benefits to athletes of endurance and weight training on muscles. (5)

....

....

....

....

....

....

Option C Neurology and behaviour

 Q8 The diagram below shows a vertical section through the human brain:

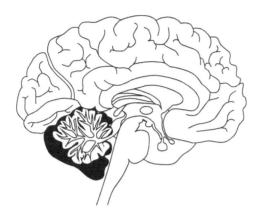

a) Label the diagram to show the position of the hypothalamus. (1)

b) Outline three functions of the hypothalamus. (3)

....

....

....

....

c) On the diagram shade the region of the brain referred to as Broca's area. (1)

d) Describe the function of Broca's area. (2)

....

....

....

....

e) A recent study by Afroz et al. (2017) demonstrated inhibition of GABA receptors in the brain triggered a reduction in the number of synapses in adolescent mice. The findings suggested that these GABA receptors are a novel target to normalise spine density (number of synapses per neurone) in adolescence and may suggest new therapies for schizophrenia where spine density and cognition are abnormal.

i) Explain the importance of synaptic pruning in adolescence. (2)

....

....

....

ii) Suggest how inhibition of GABA receptors may lead to treatments for schizophrenia, and why these findings need to be treated with caution. (3)

....

....

....

....

f) An experiment was carried out to investigate the effect of light on the behaviour of woodlice. A 'choice chamber' was made from a large Petri dish divided into two sections: light and dark, by covering half the dish with black paper. Damp filter paper was placed in each section to create a damp environment. Five woodlice were placed into the centre of the Petri dish and left for five minutes. The number of woodlice in each section was recorded every 30 seconds for four minutes, and the results are shown below. After two minutes the Petri dish was rotated 180°.

Environment	Number of woodlice found								
	0 seconds	30 seconds	60 seconds	90 seconds	120 seconds	150 seconds	180 seconds	210 seconds	240 seconds
Light and damp	2	1	1	1	0	0	2	3	2
Dark and damp	3	4	4	4	5	5	3	4	4

Using your knowledge and the table above, answer the following questions.

i) Explain why the woodlice were left for five minutes, and the dish rotated halfway through. (2)

....

....

....

ii) Name the type of innate behaviour shown. Explain your answer. (2)

....

....

....

iii) Suggest *two* reasons for the results seen after 210 seconds. (2)

....

....

....

iv) Suggest *two* improvements to the experiment to improve reliability. (2)

....

....

....

Answers

Practice questions Unit 3

3.1 ATP and 3.3 Respiration

Q1 a) Ribose sugar attached to adenine and three phosphate groups

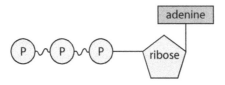

With correct labels (2)

b) Protein synthesis (1)
DNA synthesis (1)
Active transport (1)
Mitosis (1)
ANY 2
NOT muscle contraction

c) Universal energy carrier (all reactions – all organisms) (1)
Releases energy in small quantities, i.e. 30.6 kJ (1)
One-step reaction (1)

d) Substrate level phosphorylation involves transfer of phosphate groups from donor molecules (1)
e.g. glycerate-3-phosphate to ADP during glycolysis (1)
whereas oxidative phosphorylation occurs when a phosphate ion is added to ADP (1)
e.g. using energy from oxidation reactions / energy from electron loss (1)
ANY 3

Q2 a)

Statement	Glycolysis	Link reaction	Krebs cycle	Electron transport chain
Occurs in the mitochondrial matrix	✗	✓	✓	✗
ATP produced by substrate-level phosphorylation	✓	✗	✓	✗
FAD reduced	✗	✗	✓	✗
NADH$_2$ oxidised	✗	✗	✗	✓

1 mark per correct row (4)

b) ATP phosphorylates glucose (1)
producing {glucose/hexose} biphosphate / diphosphate (1)
makes molecule more reactive/easier to split (1)
into triose phosphate (1)
ANY 3

c) reduction of pyruvate to lactate (1)

regenerates NAD / oxidises $NADH_2$ (1)

Explanation
allowing glycolysis to continue (1)

lactate can be oxidised later/build-up of oxygen debt (1)

MAX 1 without explanation

Q3 a) <u>Mean</u> time taken for the solution to change colour is lowest / quickest at 45 °C indicating the fastest rate of reaction (1)

However, range/error bars overlap between 45 and 50 °C indicating one result for 50 °C was lower/faster than at 45 °C (1) allow converse

The true optimum temperature is likely to lie between 45 and 50 °C, further repeats would be needed to confirm (1)

b) Add a pH buffer (pH 7.0) to ensure pH is kept constant (1)

Use of colorimeter to determine colour change quantitatively, rather than timing which is subjective (1)

c) At 55 °C the higher temperature causes the dehydrogenase enzyme molecules to vibrate more / ref increased kinetic energy (1)

This causes the hydrogen bonds between amino acids to begin to break, changing the tertiary structure of the enzyme (1)

Fewer substrate molecules can bind to the dehydrogenase enzymes active sites, reducing the number of enzyme–substrate complexes (1)

Allow ref to denaturing NOT denatured

d) Dehydrogenase enzymes accept H^+ from oxidised molecules in link reaction and Krebs cycle (1)

passing the H^+ to NAD/FAD which are reduced to $NADH_2/FADH_2$ (acc NADH/FADH) (1)

TTC is an artificial hydrogen acceptor and so has a complementary shape and {fits into the active site of dehydrogenase / forms an enzyme–substrate complex} and is also reduced (1)

Q4 a) i) Glycolysis in cytoplasm (1)

Link, Krebs in mitochondrial matrix (1)

Electron transport chain in cristae (1)

ii) $efficiency = \dfrac{energy\ from\ ATP}{energy\ in\ glucose}$

$\dfrac{30.6 \times 38}{2880} \times 100$

= 40.4% (1 dp)

Answer correct = 2

Answer wrong but correct working shown = 1

b) Anaerobic respiration involves only glycolysis and yields just 2 net ATP molecules per glucose, 2 pyruvate and 2 NADH molecules (1)

Pyruvate is then reduced to lactate {regenerating NAD/oxidising NADH} (1)

(Anaerobic respiration cannot continue for long, and once oxygen is available) lactate is readily oxidised when it enters Krebs cycle, releasing further ATP molecules (1)

Q5 a) When a phosphate {ion/group} is added to ADP using energy from {electron loss/oxidation reactions} (1)

b) NADH {utilises three proton pumps/joins the electron transport chain at a higher energy level} (1)

so is able to pump three proton pairs from the matrix into the inter-membrane space using energy from the electrons (1)

whereas FADH only utilises two proton pumps so pumps two proton pairs (1)

Each proton pair with ATP synthetase catalyses the phosphorylation of one ADP molecule (1)

ANY 3

c) Electrons cannot pass to the terminal electron acceptor to form water so flow of electrons soon stops (1)

ATP can only be made during glycolysis, link and Krebs so ATP yield is greatly reduced (1)

NADH is no longer {regenerated / oxidised to NAD} so eventually glycolysis also stops as triose phosphate cannot be oxidised to pyruvate (1)

ANY 2

Non-competitive inhibition cannot be overcome by increasing {substrate / electron concentration} as the allosteric site of the proton pump is affected not the active site (1)

Q6

a) $2 \times 3.14 \times 0.6 \times 11.2 + 2 \times 3.14 \times 0.6^2$

$= 42.20 + 2.26$

$= 44.46 \ \mu m^2$

Answer correct = 2

Credit working if answer wrong

b) Muscles are metabolically active / need to produce high quantities of ATP via aerobic respiration (1)

{Aerobic respiration/ link, Krebs cycle, electron transport chain} occurs in the matrix/cristae of mitochondrion (1)

Pyruvate from glycolysis has to be transported into the mitochondria (1)

ANY 2 plus

Larger SA increase rate of {facilitated diffusion/active transport} of pyruvate as number of transport molecules increased (1)

Q7

a) 1 between triose phosphate and pyruvate (1)

1 between 5C and 4C in Krebs (1)

b) Fats (no mark)

Fatty acids have large numbers of carbon and hydrogen atoms (1)

Respiring them yields more carbon dioxide, water and ATP due to more hydrogen being utilised in the electron transport chain (1)

c) The excess amino acids are deaminated in the liver and the amine group NH_2, is converted to urea in the ornithine cycle (1)

The carboxyl group that remains can be converted into a number of different Krebs cycle intermediates (1)

3.2 Photosynthesis

Q1 a) Thylakoids/thylakoid membrane/granum (1)

b) Photophosphorylation (1)

c) Nucleotides (1)

d) Photolysis/splitting of water (1)
replaces electrons lost from {chlorophyll/PSII} (1)
providing {protons/H$^+$} (1)
which reduces NADP/ used for ATP synthesis (1)
ANY 3

e) Electrons fall back into electron transport chain and {take a cyclical route/ref to cyclic photophosphorylation} (1)
generating 1 ATP (1)

Q2 a) Chloroplasts can change energy from one form into another, in this case light energy into chemical energy (1)

b) The absorption spectrum is a graph which shows the quantity of light a particular pigment absorbs at each wavelength whereas the action spectrum is a graph which shows the rate of photosynthesis at different wavelengths of light (1)

c)

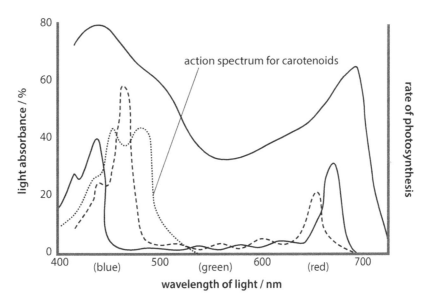

Correct line marked (1)
Explanation = carotenoids absorb light energy from the blue-violet region of the spectrum (1)

d)

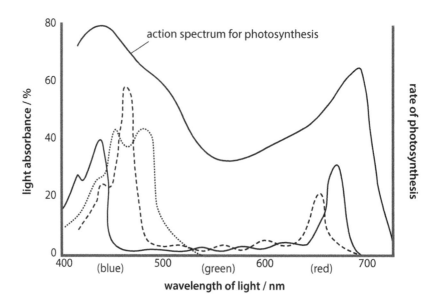

Must be clearly labelled

e) Large fronds to absorb light energy at {low light intensities at 30 m surface/ref to 12.5% light intensity} (1)

Brown colour due to high proportion of {carotenoids / chlorophyl b} pigments to absorb blue wavelength light found at 30 m (1)

Q3 a) The bacteria congregated around the algae exposed to blue and red wavelengths (1)

This is because the cells were photosynthesising and releasing oxygen which attracted the motile aerobic bacteria (1)

Very few aerobic bacteria were found near green wavelengths of light due to absence of oxygen (1)

b) An array of protein and pigment molecules in the thylakoid membranes (1)

with chlorophyll a at the reaction centre

Function = it transfers energy from light of a range of wavelengths to chlorophyll a (1)

Q4 a) Limiting factor is the one which is in shortest supply which controls the rate-limiting step, and therefore an increase in it increases the rate of photosynthesis (1)

b) X = carbon dioxide

At low concentrations carbon dioxide concentration is limiting, but above 0.5%, the rate plateaus, showing that something else must be limiting. Above about 1%, the stomata close, preventing uptake of carbon dioxide so a decrease is seen.

Y = light intensity

As light intensity increases, the rate of photosynthesis increases up to about 10,000 lux when some other factor becomes limiting. At very high light intensities the rate decreases as chloroplast pigments become bleached.

Z = temperature

Temperature increases the kinetic energy of the reactants and enzymes involved in photosynthesis. Unlike other factors, a plateau is not reached, as enzymes, e.g. RuBisCO, begin to denature so the rate of photosynthesis decreases above the optimum temperature.

Q5 You are not awarded a tick per point, but rather an assessment of your answer is made awarding three main bands. Awarding a mark within the band will depend on how fully you meet the statement.

7–9 marks

Indicative content of this level is…

Detailed comparison of ATP synthesis in both mitochondria and chloroplasts

Evaluation of similarities and differences between ATP production in both

The candidate constructs an articulate, integrated account, correctly linking relevant points, such as those in the indicative content, which shows sequential reasoning. The answer fully addresses the question with no irrelevant inclusions or significant omissions. The candidate uses scientific conventions and vocabulary appropriately and accurately.

4–6 marks

Indicative content of this level is…

Comparison of ATP synthesis in both mitochondria and chloroplasts

Some evaluation of similarities and differences between ATP production in both

The candidate constructs an account correctly linking some relevant points, such as those in the indicative content, showing some reasoning. The answer addresses the question with some omissions. The candidate usually uses scientific conventions and vocabulary appropriately and accurately.

1–3 marks

Indicative content of this level is…

Basic comparison of ATP synthesis in both mitochondria and chloroplasts

Little or no evaluation of similarities and differences between ATP production in both

The candidate makes some relevant points, such as those in the indicative content, showing limited reasoning. The answer addresses the question with significant omissions. The candidate has limited use of scientific conventions and vocabulary.

0 marks

The candidate does not make any attempt or give a relevant answer worthy of credit.

A good answer would therefore include:

Similarities
- Uses energy from electrons to pump protons across the membrane which then flows back through stalked particles
- Involves membrane-bound proton pumps
- Uses ATP synthetase

Differences
- Proton gradient is established between inter-membrane space to matrix in mitochondria whereas in chloroplasts it is from thylakoid space to stroma
- Co-enzymes involved in mitochondria are FAD and NAD, whereas in chloroplasts it is NADP
- Terminal electron acceptor is oxygen and H^+ in mitochondria whereas it is NADP and H^+ in non-cyclic photophosphorylation and chlorophyll$^+$ in cyclic photophosphorylation
- Site of electron transport chain is the cristae in mitochondria whereas in chloroplasts it is the thylakoid membrane

Evaluation
- Critical evaluation that essentially same process but with some names different, e.g. matrix vs stroma
- Some processes are different, e.g. terminal electron acceptor, co-enzymes involved

Q6 a)

Temperature / °C	Length of bubble in capillary tube / mm				Mean volume of oxygen produced in five minutes / mm³
	Trial 1	Trial 2	Trial 3	Mean	
20	25	23	22	**23.3**	**18.3**
25	32	34	40	**35.3**	**27.7**
30	41	42	42	**41.7**	**32.7**
35	45	47	40	**44.0**	**34.5**
40	32	30	30	**30.7**	**24.1**

All correct = (3), −1 for each incorrect

b) The highest mean volume of oxygen collected was 34.5 mm³ at 35 °C (1);

however, the three trials were not consistent, i.e. trial 3 produced a volume lower than collected at 30 °C / ref to 40 mm vs 41, 42 mm bubble length at 30 °C (1)

further repeats needed at 35 °C to confirm true mean (1)

3.4 Microbiology

Q1 a) Sanitation / safe disposal of sewage / good hygiene practices (1)

Provision of {clean / safe} drinking water / bottled water (1)

Use of vaccine

NOT antibiotic use/ oral rehydration therapy

b) i) Lipoprotein (1)

Lipopolysaccharide (1)

ii) Y= peptidoglycan / murein (1)

iii) Red/pink (1)

c) Penicillin prevents formation of peptidoglycan / cross-links in cell wall (1)

Cholera is a Gram-negative bacterium (1)

so has very little peptidoglycan NOT no peptidoglycan (1)

Lipopolysaccharide layer protects cell from penicillin action (1)

ANY 3

d) Serial dilution, ref to 1/10 with sterile {water/medium} (1)

Aseptic technique / ref to flaming neck of bottle / loop / sterile pipette used, etc. (1)

{1 cm³ or specified volume} spread onto sterile agar plates and incubated at 37 °C (1)

plate with 10–100 colonies chosen and counted multiplied by dilution factor (1)

Q2 a) Source of carbon for respiration (1)

Source of nitrogen for nucleotide/protein synthesis (1)

b) Tube = C (1)

Bacteria cannot survive in presence of oxygen so are only found at base of tube (1)

c) Tube = A (1)

Bacteria require oxygen for growth so are found at top of tube close to surface where oxygen is present (1)

d) Microbes that grow better with oxygen but can grow without it (1)

    e) 59 colonies (1)

       from a 10,000 dilution

       number of bacteria in 0.1 cm^3 is 59 × 10,000 = 590,000 (1)

       = 10 × 590,000 = 5.9 million per cm^3 (1)

Q3 a) Wash hands /disinfect bench (1)

       Sterilise loop by passing through Bunsen flame/use sterile pipette to transfer sample (1)

       Flame neck of culture bottle (1)

       Re-sterilise loop afterwards (1)

       ANY 2

    b) i) Gram negative (1)

          Cocci (allow coccus) (1)

      ii) Differences in cell wall structure (1)

          {purple/Gram-positive} bacteria have thicker cell wall (1)

          made of peptidoglycan/murein (1)

          which takes up {gram/purple} stain/crystal violet (1)

          {pink/Gram-negative} bacteria have lipopolysaccharide layer that does not take up stain (1)

          MAX 3

Q4 a) A = Lag phase

       B = Exponential/log/growth phase

       C = Stationary phase

       All 3 = 2 marks

       2 correct = 1 mark

    b) Bacteria used up glucose so enters stationary phase (1)

       begins synthesising enzymes to hydrolyse starch to glucose so growth can continue (1)

       Allow ref to diauxic growth (1)

       ANY 2

    c) Death phase = more cells are dying than are being produced so the population decreases (1)

       Death of cells is due to lack of nutrients, lack of oxygen or increased toxicity of the medium (1)

3.5 Population size and ecosystems

Q1 a) Any point on line between 10 and 12 hours (1)

    b) i) The maximum number around which a population fluctuates in a given environment (1)

      ii) Line fluctuates around $\log_{10} = 6$ (1)

    c) Rate of growth per day $= \dfrac{\text{antilog}_{10}5 - \text{antilog}_{10}2}{5}$

       $= \dfrac{100\,000 - 100}{5}$

       = 19 980 per day (2)

    d) Density-dependent have an increased effect on larger population sizes, e.g. disease / predation / biotic factor (1)

       Density-independent factors have the same effect regardless of population size, e.g. temperature / light intensity / abiotic factor (1)

Q2

a) A community in which energy and matter are transferred in complex interactions (1)

between the environment and organisms, involving biotic and abiotic elements (1)

b) Some of the biomass is used to form inedible material, e.g. bark, or is biomass in roots, which is out of reach of primary consumers (1)

c) Tropical rainforest has high rainfall, warm temperature and high light intensity whereas rainfall, temperature and light intensity are lower in temperate deciduous forest (1)

reducing the rate of photosynthesis and therefore amount of energy fixed by plants (1)

Most trees in a temperate deciduous forest will be {dormant/ref to loss of leaves} during winter months further reducing production per year (1)

Q3

a) $87\,000 / 1.7 \times 10^6 \times 100 = 5.1\%$ (2)

Allow 1 mark for correct process but incorrect answer

b) Able to digest high protein diets more efficiently (1)

Less energy is available 125 kJ vs 15 000 so has to be more efficient (1)

c) Some light is reflected from the leaf surface

Some light is the wrong wavelength, e.g. green, ultraviolet (1)

Some light passes through the leaf without being captured by chloroplasts (1)

3 correct = 2 marks

2 correct = 1 mark

Q4 You are not awarded a tick per point, but rather an assessment of your answer is made awarding three main bands. Awarding a mark within the band will depend on how fully you meet the statement.

7–9 marks

Indicative content of this level is...

Detailed description of primary succession including all key changes in soil, species diversity and stability of the community.

Factors affecting succession including interspecific and intraspecific competition explained.

The candidate constructs an articulate, integrated account, correctly linking relevant points, such as those in the indicative content, which shows sequential reasoning. The answer fully addresses the question with no irrelevant inclusions or significant omissions. The candidate uses scientific conventions and vocabulary appropriately and accurately.

4–6 marks

Indicative content of this level is...

Description of primary succession including main changes in soil, species diversity and stability of the community.

Some factors affecting succession, e.g. interspecific or intraspecific competition described.

The candidate constructs an account correctly linking some relevant points, such as those in the indicative content, showing some reasoning. The answer addresses the question with some omissions. The candidate usually uses scientific conventions and vocabulary appropriately and accurately.

1–3 marks

Indicative content of this level is...

Basic description of primary succession including some changes in soil, species diversity and stability of the community.

Limited explanation of factors affecting succession, e.g. interspecific or intraspecific competition.

The candidate makes some relevant points, such as those in the indicative content, showing limited reasoning. The answer addresses the question with significant omissions. The candidate has limited use of scientific conventions and vocabulary.

0 marks

The candidate does not make any attempt or give a relevant answer worthy of credit.

A good answer would therefore include:

- Weathering creates small cracks in the rocks and small particles.
- Mosses and lichens begin to colonise. Organic matter builds up slowly.
- Legumes begin to grow as they are able to fix atmospheric nitrogen to supplement the poor nutrient soil. As they die, soil becomes enriched.
- Grasses and ferns start to grow, sheltering the soil from the elements. Soil, and its moisture content increases.
- Large shrubs and small trees colonise. Leaf litter greatly increases fertility and humus content of the soil. Habitats created for nesting birds and soil invertebrates so diversity increases.
- Climax woodland is reached. This is usually oak, beech, hazel or lime species but is largely deciduous in Southern UK. Ground flora includes bracken, shrubs and bluebell.

Changes:

- Soil depth increases
- Nutrient content increases
- Humus content increases therefore water content increases
- Species diversity increases
- Stability of community increases.

As new species are introduced, competition exists for resources at all the seral stages because, for example, legumes can outcompete mosses as the soil content increases. Competition exists between:

1. Different species (interspecific competition) where each may occupy a different niche.
2. Individuals of the same species (intraspecific competition) which is density dependent, i.e. competition increases with population size.

Q5 a) 1 mark for each axis correctly labelled (2)

1 mark for each line correctly plotted (3)

MAX 3

b) Ammonium ions decrease from 7 to 1 mg dm^{-3} as nitrification occurs (1)

Ammonium ions are converted to nitrite (1)

by bacteria, e.g. Nitrosomonas (1)

c) Between 6 and 12 days nitrite levels increase from 1 to 10 mg dm^{-3} and then decrease back to 1 by day 21 (1)

Increases as ammonium ions converted to nitrite (by Nitrosomonas bacteria) and decreases as nitrite converted to nitrate by Nitrobacter bacteria (1)

d) {On/after} day 18 (1)

Nitrate levels begin to fall as nitrate is taken up by plant (1)

and converted into protein (or other named nitrogen compound, e.g. nucleotide) in plant (1)

Q6 a) A = Combustion

B = Respiration

C = Assimilation

D = Exposure and erosion

All 4 correct = 3 marks

3 correct = 2 marks

2 correct = 1 mark

b) Carbon dioxide is dissolved in aquatic ecosystems as HCO^{3-} ions, and forms calcium carbonate in mollusc shells and arthropod skeletons (1)

When these organisms die, and their shells settle on the ocean bed (1)

compression over millions of years forms {chalk/limestone /marble}, from these carbonates (1)

3.6 Human impact on the environment

 Q1 You are not awarded a tick per point, but rather an assessment of your answer is made awarding three main bands. Awarding a mark within the band will depend on how fully you meet the statement.

7–9 marks

Indicative content of this level is...

Detailed explanation of reasons why numbers have declined and how trend could be reversed. Good use of data from graphs to support argument.

The candidate constructs an articulate, integrated account, correctly linking relevant points, such as those in the indicative content, which shows sequential reasoning. The answer fully addresses the question with no irrelevant inclusions or significant omissions. The candidate uses scientific conventions and vocabulary appropriately and accurately.

4–6 marks

Indicative content of this level is...

Explanation of reasons why numbers have declined and how trend could be reversed. Some use of data from graphs to support argument.

The candidate constructs an account correctly linking some relevant points, such as those in the indicative content, showing some reasoning. The answer addresses the question with some omissions. The candidate usually uses scientific conventions and vocabulary appropriately and accurately.

1–3 marks

Indicative content of this level is...

Basic description of reasons why numbers have declined and how trend could be reversed. Little or no use of data from graphs to support argument.

The candidate makes some relevant points, such as those in the indicative content, showing limited reasoning. The answer addresses the question with significant omissions. The candidate has limited use of scientific conventions and vocabulary.

0 marks

The candidate does not make any attempt or give a relevant answer worthy of credit.

A good answer would therefore include:

Reasons for decline:

- Logging being primary threat to habitat
- Deforestation definition – cutting down trees and use of land for another purpose
- Rainforest wildfires
- Growing of cash crops, e.g. palm oil, to meet needs of biofuels, foods, cosmetics
- Poaching for bush meat and traditional medicines
- Supplying pet trade
- Use of data from graphs, e.g. % increase in palm oil production

Conservation methods:

- CITES
- EU habitats directive
- Establish protected areas
- Captive breeding programmes
- Sperm banks
- Role of zoos in conservation projects and breeding programmes
- Species reintroduction
- Education
- Ecotourism

a) They remove carbon dioxide from the atmosphere during photosynthesis (1)

They are not completely carbon-neutral as energy is used in their production, processing and distribution (1)

b) Deforestation has led to mono-culture of biofuel crops resulting in land system change boundary being crossed (1)

and biodiversity integrity boundary being crossed due to loss of species (1)

c) Using water-efficient appliances (1)

Reclaiming waste water for irrigation and industrial use (1)

Stop irrigating non-food crops (1)

Irrigate crops by using drip-irrigation systems (1)

Capture storm water run-off for recharging reservoirs (1)

Desalinate salt water (1)

Three correct = 2

3.7 Homeostasis and the kidney

 a)

Letter	Name	Function
A	**Bowman's capsule**	**Ultrafiltration**
B	**Glomerulus**	
C	**Proximal convoluted tubule**	**Selective reabsorption**
D	**Distal convoluted tubule**	Control of blood pH
E	**Loop of Henle**	**Osmoregulation**
F	Vasa recta	
G	**Collecting duct**	

All three functions correct = 2, two functions correct = 1

All six parts correctly named = 5, 5 parts = 4, 3 parts = 2, 2 parts = 1

MAX = 7

b) Medulla (1)

c) Water potential is the tendency for water to move into a system from a high water potential to a low water potential (1)

Plus any 5 from below:

Sodium ions are pumped out of the ascending limb (1)

creating low water potential in the medulla (1)

the descending limb is impermeable to sodium ions but permeable to water (1)

water moves out of descending limb AND collecting duct (1)

by osmosis (1)

ref to counter current (1)

d) Desert animals need to conserve water so produce highly concentrated urine (1)

by reabsorbing more water by osmosis (1)

due to higher solute potential gradient created by longer {loop of Henle / E} (1)

by more sodium ions being actively transported out of ascending limb into medulla (1)

MAX 3

e) ADH binds to membrane receptor proteins found on the surface of cells lining the duct (1)

ADH binding triggers vesicles containing intrinsic membrane proteins called aquaporins to fuse with the cell membrane (1)

The aquaporins contain pores allowing water to move (1)

The higher the concentration of ADH the more aquaporins fuse with the membrane (1)

ANY 3

Q2 a) A   Proximal convoluted tubule

     B   Bowman's capsule

     C   Glomerulus

     D   Basement membrane

     4 correct = 3, 3 correct = 2, 2 correct = 1

b) Cortex (1)

c) Selective reabsorption (1)

Adaptations – {microvilli/folded base membrane/basal channels} to increase surface area for absorption (1)

Large no. of mitochondria provide ATP for active transport of, e.g. {glucose/amino acids} (1)

Q3 a) Medulla (1)

Reason – Bowman's capsules not visible / {collecting duct/loop of Henle} visible (1)

b) e.g. 9 mm (allow correct measurement from diagram) (1)

width = size of image / magnification

= 9/400 = 0.023 mm × 1000

= 23 mm (1)

−1 no units

c) Function = absorption (1)

{Basement membrane \D} is thin to reduce diffusion distance (1)

Allow ref to sodium/chloride channels creating decreased water potential in medulla

Q4 You are not awarded a tick per point, but rather an assessment of your answer is made awarding three main bands. Awarding a mark within the band will depend on how fully you meet the statement.

7–9 marks

Indicative content of this level is...

Detailed explanation of how each part of the nephron is adapted for all of its functions including ultrafiltration, selective reabsorption, osmoregulation.

The candidate constructs an articulate, integrated account, correctly linking relevant points, such as those in the indicative content, which shows sequential reasoning. The answer fully addresses the question with no irrelevant inclusions or significant omissions. The candidate uses scientific conventions and vocabulary appropriately and accurately.

4–6 marks

Indicative content of this level is...

Explanation of how each part of the nephron is adapted for most of its functions including ultrafiltration, selective reabsorption, osmoregulation

The candidate constructs an account correctly linking some relevant points, such as those in the indicative content, showing some reasoning. The answer addresses the question with some omissions. The candidate usually uses scientific conventions and vocabulary appropriately and accurately.

1–3 marks

Indicative content of this level is...

Basic explanation of how some parts of the nephron are adapted for their functions.

The candidate makes some relevant points, such as those in the indicative content, showing limited reasoning. The answer addresses the question with significant omissions. The candidate has limited use of scientific conventions and vocabulary.

0 marks

The candidate does not make any attempt or give a relevant answer worthy of credit.

A good answer would therefore include:

Ultrafiltration:

- Ref to wider afferent vessel increasing blood pressure in glomerulus
- Pores in endothelial cells / basement membrane and podocytes act as a molecular filter allowing molecules <70,000 rmm to pass into filtrate/capsule
- Example of substance which passes out, e.g. glucose, amino acids and one which does not, e.g. large proteins, cells

Selective reabsorption:

- Proximal convoluted tubule cells have microvilli/folded base membrane / basal channels to increase surface area for absorption
- Large numbers of mitochondria to provide ATP for active transport of glucose/amino acids

Osmoregulation:

- Ascending loop of Henle actively transports sodium ions OUT but is impermeable to water which decreases water potential in medulla region
- Descending limb is impermeable to water, so water passes out by osmosis
- ADH affects receptors in collecting duct walls / distal convoluted tubule making them more permeable to water and so more water is absorbed by osmosis making urine more concentrated

Q5

a) Excess amino acids are deaminated in the liver (1)

to produce ammonia and organic acid (1)

b) Ammonia is a small, very soluble but is highly toxic so must be excreted immediately (1)

It cannot be stored and requires large volumes of water which are present in the freshwater environment to dilute it down to non-toxic levels for it to be excreted safely (1)

c) Uric acid is virtually non-toxic and can therefore be stored for long periods of time (1)

Very little water is needed to safely excrete it which is an advantage to flying birds as it reduces weight (1)

It allows these animals to survive in very dry environments (1)

Q6

a) hCG must be <70,000 RMM to form part of glomerular filtrate (1)

hCG is not selectively reabsorbed in proximal convoluted tubule so remains in urine (1)

b) Glucose is <70,000 RMM so is part of glomerular filtrate (1)

Allow marking point in either part a or part b but not both

Blood glucose levels are so high in diabetic patients that not all the glucose can be selectively reabsorbed in the proximal convoluted tubule, some remains (1)

In healthy individuals, all the glucose is reabsorbed in proximal convoluted tubule by secondary active transport (1)

MAX 2

c) Urea is forced out by ultrafiltration / forms part of glomerular filtrate (1)

Urea is not selectively reabsorbed, but water is so {proportion of urea to water decreases / concentration of urea increases} (1)

3.8 The nervous system

Q1

a) Motor (neurone) (1)

b) From left to right from cell body (1)

c) A = Nucleus

B = Cell body

C = Axon

D = Myelin sheath

All 4 = 3 marks, 3 correct = 2 marks, 2 correct = 1 mark

d) Myelination (1)

Depolarisation only occurs at nodes of Ranvier, allowing action potential to 'jump' from node to node speeding up impulse (1)

Diameter of axon (1)

Larger diameter axons have less leakage of ions so increase speed of transmission (1)

NOT temperature as mammals maintain constant internal body temperature (this is a mammalian neurone)

Q2

a) Resting potential (1)

b) Membrane is more permeable to potassium ions / impermeable to sodium ions (1)

Some potassium ion gates are open allowing potassium ions to pass out (1)

Sodium ion gates are closed preventing sodium ions entering (1)

Sodium potassium ion gates transport three sodium ions out for every two potassium ions in (1)

Resulting in the inside being less positive than outside (1)

ANY 3

c) Period marked after 3 ms peak to establishment of resting potential after 6 ms (1)

d) Arrow drawn from right to left (1)

e) Relative refractory period is the period during which it is possible to send another impulse even if the stimulus is big enough to overcome threshold (1)

Whereas the absolute refractory period is the period during which it is NOT possible to send another impulse, irrespective how BIG the stimulus is (1)

Need comparison

f) Neurone is polarised / –65 mv (1)

Stimulus arrives causing Na^+ gates to OPEN (1)

Na^+ ions rush in, depolarising the neurone (1)

Now charge across membrane becomes MORE positive (MORE positive charges inside) (1)

As more Na^+ ions enter, more gates open so even more Na^+ ions rush in until potential reaches +40 mv (*positive feedback*) (1)

neurone is said to be depolarised (1)

ANY 5

Q3

a) Calcium ions flood into pre-synaptic knob following arrival of impulse (1)

causing vesicles containing acetylcholine to migrate and fuse with the pre-synaptic membrane (1)

resulting in acetylcholine being released into synaptic cleft by exocytosis (1)

b) i) Excitatory / stimulant (1)

Reason = increasing concentration of methamphetamine causes an increase in acetylcholine release (1)

Ref to twofold increase in acetylcholine release / % increase from graph / ref to numbers, e.g. increases from 100% to 250–300 % (1)

ii) Confidence for 0 and 1 mg/kg are high due to small range/error bars (1);

however, there is a large degree of variance in 4 mg/kg results suggesting effect could be higher (1)

only nine rats were tested, or conditions they were kept in, which reduces confidence in conclusion (1)

c) Results show increasing concentrations of nicotine cause a decrease in acetylcholine being released / ref to values (1)

making the person reliant on nicotine to act as a neurotransmitter (1)

Q4 You are not awarded a tick per point, but rather an assessment of your answer is made awarding three main bands. Awarding a mark within the band will depend on how fully you meet the statement.

7–9 marks

Indicative content of this level is...

Detailed description of synaptic transmission and detailed explanation of how organophosphorus insecticides affect synapses.

The candidate constructs an articulate, integrated account, correctly linking relevant points, such as those in the indicative content, which shows sequential reasoning. The answer fully addresses the question with no irrelevant inclusions or significant omissions. The candidate uses scientific conventions and vocabulary appropriately and accurately.

4–6 marks

Indicative content of this level is...

Description of synaptic transmission and explanation of how organophosphorus insecticides affect synapses.

The candidate constructs an account correctly linking some relevant points, such as those in the indicative content, showing some reasoning. The answer addresses the question with some omissions. The candidate usually uses scientific conventions and vocabulary appropriately and accurately.

1–3 marks

Indicative content of this level is...

Basic description of synaptic transmission and limited explanation of how organophosphorus insecticides affect synapses.

The candidate makes some relevant points, such as those in the indicative content, showing limited reasoning. The answer addresses the question with significant omissions. The candidate has limited use of scientific conventions and vocabulary.

0 marks

The candidate does not make any attempt or give a relevant answer worthy of credit.

A good answer would therefore include:

Synaptic transmission:

- Calcium channels open and ions flood into synaptic knob
- Synaptic vesicles migrate to and fuse with the pre-synaptic membrane
- Neurotransmitter / acetylcholine released into cleft
- Diffuses across cleft and binds to receptors on the post-synaptic membrane
- Causing sodium channels to open
- Sodium ions flood in, depolarising post-synaptic neurone

Effect of organophosphorus insecticides:

- Inhibit cholinesterase enzymes
- Causing acetylcholine to remain bound to receptors on post-synaptic neurone
- Causing repeated depolarisation of post-synaptic neurone

Q5 a)

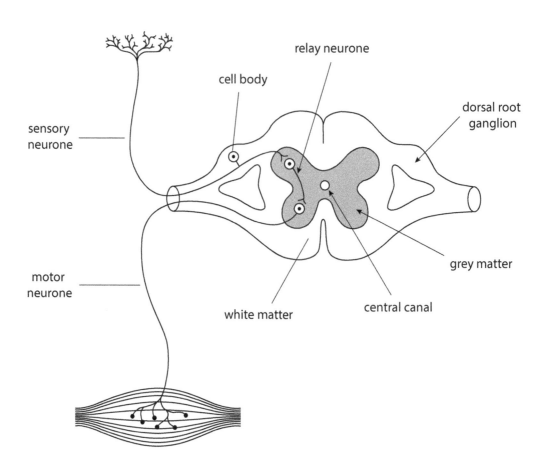

All 5 = 4 marks, −1 for each incorrect

b) Heat/stimulus detected by receptor in skin (1)
Impulse/action potential travels to CNS via sensory neurone (1)
Synapses with connector/relay neurone (1)
Relay of impulse/action potential to brain (1)
And motor neurone to the effector muscle (1)
Which brings about a response to contract muscle to withdraw hand (1)
ANY 5

Practice questions Unit 4

4.1 Sexual reproduction in humans

Q1 a) B = Primary spermatocyte
C = Secondary spermatocyte
D = Spermatid
3 correct = 2 marks
2 correct = 1 mark

b) Cell B is diploid and undergoes meiosis to produce (1)
Cell C which is haploid (1)

c) Cell D undergoes differentiation (1)
and incorporates an acrosome in the head of the spermatozoa (1)
which contains hydrolytic enzymes allowing it to digest zona pellucida of ovum (1)
A mid piece is added containing numerous mitochondria producing ATP for locomotion (1)
and a tail which provides movement towards the secondary oocyte (1)
ANY 4

Q2 a) A = Seminal vesicle
B = Prostate gland
C = vas deferens
D = Epididymis
E = Seminiferous tubules
All 5 = 3 marks
4 = 2 marks
3 = 1 mark

b) A = Produces secretion to aid sperm motility (1)
B = Produces alkaline secretion to neutralise acidity of urine (1)

c)

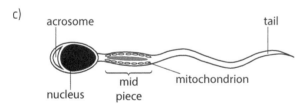

1 mark for diagram
2 marks for correct labels

Q3 a) A = Progesterone (1)
levels remain high from day 20 to inhibit FSH and LH (1)
B = FSH/follicle stimulating hormone (1)
peaks around day 5 to promote maturation of Graafian follicle
C = LH/Luteinising hormone (1)
peaks around day 14 to stimulate ovulation (1)
D = Oestrogen (1)
increases from day 4 to 14 to increase the thickness and vascularity of the uterus lining (1)

b) Anterior pituitary gland (1)

c) Oestrogen (1) as it stimulates LH (1)

d) Progesterone (1) as it would inhibit FSH, so no oestrogen produced, resulting in no stimulation of LH production (1)

Q4 a) Acrosome reaction is where acrosome enzymes digest the zona pellucida allowing sperm and oocyte membranes to fuse (1)

whereas the cortical reaction occurs when the cortical granule membranes fuse with the oocyte membrane converting it into a fertilisation membrane (1)

Acrosome reaction assists in entry of spermatozoon, whereas cortical reaction prevents entry of further spermatozoa (1)

ANY 2

b)

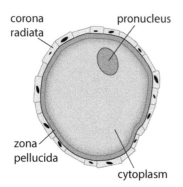

c) Produces hormones to maintain pregnancy (1)

Acts as a physical barrier reducing blood pressure between maternal and foetal circulations / separates immune system preventing maternal immune response (1)

Allows antibodies to cross placenta giving some passive immunity (1)

Removes waste from foetal blood, e.g. CO_2 (1)

NOT supplies nutrients

ANY 3

Q5 a) A = Corpus luteum

B = Secondary oocyte

C = Graafian follicle

All 3 = 2 marks,

2 correct = 1 mark

b) Both are produced by meiosis I/first meiotic division (1)

Secondary spermatocyte is produced in the testis shortly after the primary spermatocyte (1)

whereas the secondary oocyte is produced in the ovary prior to ovulation (1)

One primary spermatocyte gives rise to four secondary spermatocytes (1)

whereas one primary oocyte gives rise to one secondary oocyte and one polar body (1)

Need comparison

MAX 4

4.2 Sexual reproduction in plants

 a)

Insect-pollinated flower	Wind-pollinated flower
Large brightly coloured petals to attract insects (do not allow ref to nectar / scent as not shown)	Petals absent, as no need to attract insects
Anthers inside flower which transfer pollen to insects when they feed on nectar	Anthers hanging outside the flower so wind can blow pollen away
Stigma within the flower to collect pollen from insect when it feeds on nectar	Large feathery stigmas providing a large surface area for wind to catch pollen grains

Must be comparative

ANY 3 matched statements

b) Small quantities of sticky sculptured pollen to stick to insect (1)

Large quantities of small, smooth, light pollen to be carried by wind (1)

 a) Pollination is the transfer of pollen from the anther of one flower to the mature stigma of another flower of the same species, whereas fertilisation is the process where the male gamete fuses with the female gamete, producing a diploid zygote.

b) The pollen grain germinates producing a pollen tube (1)

The growth of the tube is controlled by the pollen tube nucleus, which also produces hydrolases, e.g. cellulases and proteases which digest a path through the style towards the micropyle (1)

This is guided by chemical attractants, e.g. GABA. (1)

The tube nucleus then disintegrates, and the two male gametes enter the ovule (1)

ANY 3

c) i) To maintain a humid environment/prevent slides from drying out (1)

ii) Coverslip might have affected growth of pollen tube (1)

iii) 0.4 mol dm^{-1} produced highest germination at 70% and longest mean tube length of 290 μm (1)

10 repeats were sufficient to produce reliable results (1)

concentrations used did not increase by the same amounts / ref to the mean could lie between 0.4 and 0.8 mol dm^{-1} so further repeats needed, e.g. at 0.6 mol dm^{-1} (1)

 a) Increase = 16 mm (1)

16/22 × 100 = 72.7% accept 73% (1)

b) Gibberellic acid increased mean height of seedlings by 73% (16 mm) compared to the control at 20 days/10 days after application / gibberellic acid increased growth in pea seedlings (1)

Gibberellic acid is a plant growth regulator which switches on genes involved in transcription and translation, resulting in the production of enzymes responsible for growth (1)

c) Only two plants in each group were measured, which reduces confidence in results (1)

repeat with more plants, e.g. 10 (1)

4.3 Inheritance

Q1

a) Gene: a sequence of DNA occupying a specific locus on a chromosome normally coding for a specific polypeptide (1) whereas an allele is a different form of the same gene (1)

b) Homozygous is where both alleles for a given characteristic are the same, whereas (1) Heterozygous is where both alleles for a given characteristic are different (1)

c) Dominant is where alleles are always expressed, i.e. in both homozygote and heterozygote, e.g. RR or Rr whereas (1) recessive is where alleles are only expressed in the homozygote, e.g. rr (1)

d) Sex linkage is where the genes are only carried on the sex chromosomes (1) whereas autosomal linkage occurs when two different genes are found on the same autosomal chromosome (non sex chromosome) and therefore cannot segregate independently (1)

Q2

a) Y = yellow, y = green, W = wrinkled, w = smooth (1)
Allow other letters if key given
Parental genotypes YyWw x yyww (1)
Gametes = YW, Yw, yW, yw x yw (1)
Offspring YyWw, Yyww, yyWw, yyww (1)
Ratio = 1 yellow wrinkled : 1 yellow smooth : 1 green wrinkled : 1 green smooth (1)

b) The characteristics of an organism are determined by factors (alleles) which occur in pairs (1)
Only one of a pair of factors (alleles) can be present in a single gamete (1)

Q3

a)

Phenotype	yellow round peas		yellow round peas	
Genotype	YyRr		YyRr	
Gametes	YR	Yr	yR	yr
YR	YYRR yellow round	YYRr yellow round	YyRR yellow round	YyRr yellow round
Yr	YYRr yellow round	YYrr yellow wrinkled	YyRr yellow round	Yyrr yellow wrinkled
yR	YyRR yellow round	YyRr yellow round	yyRR green round	yyRr green round
yr	YyRr yellow round	Yyrr yellow wrinkled	yyRr green round	yyrr green wrinkled

Phenotype ratio

9 yellow round : 3 yellow wrinkled : 3 green round : 1 green wrinkled

Gametes correct (1)
Punnet square with genotypes and phenotypes all correct (3) –1 for each error
Phenotype ratio correct (1)

b) The colour of the seed was inherited independently from the seed texture (1)
so 'each member of an allelic pair may combine randomly with either of another pair'

c) Parental genotype YyRr x YyRr
Gametes YR x yr (1)
Offspring all YyRr yellow, round (1)

d) Random mutation (1)
causing a change in the DNA / chromosome (1)
or by crossing over (1)
but only if both genes are not too close to each other (1)

Q4

a) Black coat (B) albino (b) rough coat (R) smooth coat (r) (1)

Parents BbRr x bbrr (1)

Gametes BR, Br, bR, br x br (1)

Correct offspring in punnet square (1)

Ratio 1 black rough coat 0:1 black smooth coat : 1 albino rough coat : 1 albino smooth coat (1)

b)

Category	O	E	O−E	(O−E)^2	$\frac{(O-E)^2}{E}$
black rough coat	27	25	2	4	0.16
black smooth coat	22	25	3	9	0.36
albino rough coat	28	25	3	9	0.36
albino smooth coat	23	25	2	4	0.16
	Σ = 100				Σ = 1.04

$\chi^2 = 1.04$

O−E column correct (1)

(O−E)^2 column correct (1)

$\chi^2 = 1.04$ (1)

Table (critical value at p = 0.05) = 7.82 (1)

Answer

Because the calculated value is less than the critical value at p = 0.05 (1.04 < 7.82) (1)

we can accept the null hypothesis, any difference seen was due to chance (1)

Q5

a) Pedigree diagram (1)

b) Male normal, female carrier (1)

if male was a haemophiliac, then child 7 would not be normal (would be at least a carrier as she would inherit allele from father) (1)

If female was normal then child 5 would not be a haemophiliac (as a male, the X chromosome comes from the mother) (1)

Allow converse

c) Father X^{H}y (healthy), Mother X^{H}X^h (carrier) (1)

Gametes X^H, y, X^H, X^h (1)

Offspring X^{H}y, X^{h}y (the only two male combinations) (1)

Answer 50 : 50 or 50% (1)

Q6

a) Because the related mutation is recessive, DMD is more common in boys than in girls, as boys do not have another copy of the X chromosome to compensate for the genetic defect.

b) Carrier female (1)

Male child 9 inherits X chromosome only from parent 4 so must have one copy of {faulty allele/DMD gene} as child 9 suffers from DMD (1)

c) Key X^d = DMD, X^D = normal (1)

Parent genotypes = X^d Y x X^D X^D (1)

Offspring genotypes = X^d Y, X^d Y, X^D X^d, X^D X^d (1)

(all female offspring are carriers)

Chance = 100% or 1 in 1 (1)

Unit 4 Answers

4.4 Variation and evolution

Q1 a) Variation controlled by one gene with two or more alleles.

b) There are distinct groups with no intermediates, e.g. A, B, AB and O (1)
NOT environmental factors as graph does not show this

c) Gene (point) mutations (1)
Crossing over during prophase I of meiosis (1)
Independent assortment during metaphase I and II of meiosis (1)
Random mating, i.e. that any organism can mate with another (1)
Random fusion of gametes, i.e. the fertilisation of any male gamete with any female gamete (1)
Environmental factors leading to epigenetic modifications (1)
Environmental factors can also lead to non-heritable variation within a population, e.g. diet (1)
ANY 4

Q2 a) Intraspecific competition is where members of the <u>same</u> species vie for the same resource in an ecosystem (e.g. food, light, nutrients, space) (1)
whereas interspecific competition is where individuals of <u>different</u> species vie for the same resource in an ecosystem (1)

b) The allele frequency is the relative proportions of alleles in the population, (1)
whereas the gene pool is the complete set of unique alleles in a species or population (1)

c) A species is a group of individuals with <u>similar characteristics</u> that can <u>interbreed</u> to produce <u>fertile offspring</u> (1)
whereas speciation is the evolution of new species from existing ones (1)

d) Allopatric speciation involves geographical isolation separating two populations (1)
whereas sympatric speciation involves populations living together becoming reproductively isolated by other means than a geographical barrier (1)

Q3 a) q^2 = 1 in 200 or 0.005
q = square root of 0.005 = 0.071 (1)
p + q = 1,
p = 1 – 0.071 = 0.929 (1)
frequency of heterozygotes = 2pq = 2 × 0.929 × 0.071 = 0.132 (1)
= 0.132 × 100 = 13.2% / or 1 in 7.576 are carriers (1)

b) p^2 = 0.929^2 = 0.863 (1)
0.863 × 10,000 = 8,630 (1)

c) Organisms are diploid / have equal allele frequencies in both sexes (1)
Reproduce sexually / mating is random / generations don't overlap (1)
The population size is very large / and there is no migration, mutation or selection (1)

d) Exposure to drought would confer a selective advantage for presence of the allele for a thick cuticle /t (1)
Only homozygous recessive plants (tt) would produce a thick cuticle and survive / plants heterozygous for cuticle thickness (TT, Tt) would die (1)
Exposure to disease would confer a selective advantage for presence of the allele for synthesis of phytoalexins / p (1)
Only homozygous recessive plants (pp) would produce phytoalexins and survive / plants heterozygous (PP, Pp) would die (1)
Frequency of dominant alleles would decrease (1)
Ref to founder effect: the loss of genetic variation in a new population established by a very small number of individuals from a larger population (1)
ANY 4

Q4

You are not awarded a tick per point, but rather an assessment of your answer is made awarding three main bands. Awarding a mark within the band will depend on how fully you meet the statement.

7–9 marks

Indicative content of this level is...

Detailed description of all mechanisms leading to development of new species, including allopatric and sympatric speciation, and adaptive radiation / founder effect.

The candidate constructs an articulate, integrated account, correctly linking relevant points, such as those in the indicative content, which shows sequential reasoning. The answer fully addresses the question with no irrelevant inclusions or significant omissions. The candidate uses scientific conventions and vocabulary appropriately and accurately.

4–6 marks

Indicative content of this level is...

Description of most mechanisms leading to development of new species, e.g. allopatric and sympatric speciation, and adaptive radiation / founder effect.

The candidate constructs an account correctly linking some relevant points, such as those in the indicative content, showing some reasoning. The answer addresses the question with some omissions. The candidate usually uses scientific conventions and vocabulary appropriately and accurately.

1–3 marks

Indicative content of this level is...

Basic description of some mechanisms leading to development of new species, e.g. allopatric and sympatric speciation, and adaptive radiation / founder effect.

The candidate makes some relevant points, such as those in the indicative content, showing limited reasoning. The answer addresses the question with significant omissions. The candidate has limited use of scientific conventions and vocabulary.

0 marks

The candidate does not make any attempt or give a relevant answer worthy of credit.

A good answer would therefore include:

- Speciation is the formation of a new species from pre-existing ones
- A species is a group of organisms with similar characteristics that can interbreed
- To produce fertile offspring
- Ref allopatric speciation
- Populations separated by a physical barrier, e.g. mountain, river
- Each population has different gene pools
- Random mutations occur
- Different mutations occur in each population
- Different selection pressures exist in each area
- Causes changes in each gene pool / allele frequency in each population
- Ref to adaptive radiation and founder effect
- As organisms adapt to different environmental conditions
- If barrier is removed, two populations are unable to breed (so are new species)
- Ref to sympatric isolation non-geographical isolation
- E.g. behavioural isolation occurs in animals with elaborate courtship behaviours where members of a sub-species fail to attract the necessary response, e.g. sticklebacks
- E.g. seasonal (temporal) isolation where organisms are isolated due to reproductive cycles not coinciding and so are fertile at different times of the year. This is seen in frogs where each of four types has a different breeding season, e.g. wood frog, pickerel frog, tree frog and bullfrog
- E.g. mechanical isolation as a result of incompatible genitalia
- E.g. gametic isolation from the failure of pollen grains to germinate on stigma or sperm fail to survive in oviduct
- E.g. failure of chromosome pairing, ref to infertile hybrids

Q5 a) Evolution is the process by which new species are formed from pre-existing ones over a long period of time. (1)

b) Horse

Environment of horse changed significantly over past 55 million years from forested area to savanna (1)

due to changes in climate / drying climate resulting in loss of trees (1)

Horse evolved longer legs / hooves for running / disadvantage to being small (1)

due to loss of tree cover / need to escape from predators (1)

ANY 3

Horseshoe crab

Very little change in anatomy seen (1)

Aquatic environment did not change much so no additional selection pressures /no need to evolve (1)

Q6 a) Despite having similar characteristics, to be regarded as the same species, the horse and donkey must be able to interbreed and produce fertile offspring (1)

A mule has a chromosome number of 63 (1)

Chromosomes fail to pair during prophase I of meiosis (1)

so gametes do not form (1)

The resulting offspring are sterile / hybrid inviability (1)

b) Sympatric speciation (1)

Populations living together becoming reproductively isolated (1)

4.5 Application of reproduction and genetics

Q1 a) Fragment size is 52 bp, so drawn below 100 bp, about ¾ up from bottom (1)

(incorrect if halfway to bottom given spacings seen above)

b) +ve at bottom, -ve at top (1)

DNA is negatively charged due to phosphate groups so is attracted to +ve electrode (1)

Smaller bands move further through the gel (1)

c) Father 1 (1)

Combination of both mother's and father's genetic fingerprint match the child's (1)

Ref to top band in child's fingerprint can only come from father 1 (1)

NOT exact match

d) i) Polymerase Chain Reaction (1)

ii) 95 °C breaks hydrogen bonds allowing DNA strands to separate (1)

50-60 °C allows primers to attach/anneal by complementary base pairing (1)

70 °C allows DNA polymerase to join complementary nucleotides / extension (1)

All three temperatures correct matched to explanation = 2 marks, one incorrect = 1 mark

MAX 5

Q2 a) 7 fragments (1)

b) 1768 – 1350 (1)

= 418 bp (1)

c) 2105 – 1768

= 337 (1)

and 1768 (1)

d) Bal I and Sna I (1)

Q3 a) Point mutation (1)

b) Adenine = purine, thymine = pyrimidine (1)

c) People who have one copy of the faulty allele do have symptoms (1)

These are not as severe as sufferers, but show increased resistance to malaria (the advantage) (1)

d) Identify healthy gene (1)

Extract healthy haemoglobin mRNA (1)

Use reverse transcriptase to produce cDNA from mRNA template (1)

Insert into plasmid / virus (1)

Inject into bone marrow (1)

ANY 4

Q4 a) Gel electrophoresis separates DNA fragments according to size (1)

Smaller fragments travel further through the gel so the smallest fragments are at the bottom / smaller fragments represent nucleotides earlier in sequence (1)

b) AGCT AGCC CCGG TAGA CC

All correct = 2

1 incorrect = 1

1 mark if sequence reversed

c) Any contamination is quickly amplified /copied (1)

DNA polymerase can sometimes incorporate the incorrect nucleotide (1)

Only small fragments can be copied (up to a few thousand bases) at a time (1)

The efficiency of the reaction decreases after about 20 cycles, as the concentrations of reagents reduce, and product builds up (1)

ANY 3

d) Primers with a higher proportion of adenine and thymine from fewer hydrogen bonds with their complementary base / ref to 1 hydrogen bond vs 2 for guanine and cytosine (1) at {60 °C/higher temperatures} hydrogen bonds would break/be unable to form between complementary bases (1)

primers unable to attach/anneal (1)

e) If a patient has a genetic predisposition to a particular disease, should this information be passed to life or health insurance companies, which could affect insurance premiums / availability of cover (1)

If ancestral relationships are determined, this could be used to socially discriminate against people (1)

If genetic diseases are identified, this has an implication for the parents and children of those diagnosed / If children are screened, when should they be told if they have a predisposition say for Alzheimer's disease? (1)

Screening of embryos could be extended from genetic diseases to desirable traits / lead to designer babies (1)

Patient data may not be stored safely (1)

ANY 3

Q5 You are not awarded a tick per point, but rather an assessment of your answer is made awarding three main bands. Awarding a mark within the band will depend on how fully you meet the statement.

7–9 marks

Indicative content of this level is...

Detailed description of the pros, cons and hazards of genetically engineering bacteria

The candidate constructs an articulate, integrated account, correctly linking relevant points, such as those in the indicative content, which shows sequential reasoning. The answer fully addresses the question with no irrelevant inclusions or significant omissions. The candidate uses scientific conventions and vocabulary appropriately and accurately.

4–6 marks

Indicative content of this level is...

Description of the major pros, cons and hazards of genetically engineering bacteria.

The candidate constructs an account correctly linking some relevant points, such as those in the indicative content, showing some reasoning. The answer addresses the question with some omissions. The candidate usually uses scientific conventions and vocabulary appropriately and accurately.

1–3 marks

Indicative content of this level is...

Basic description of some of the pros, cons and/or hazards of genetically engineering bacteria.

The candidate makes some relevant points, such as those in the indicative content, showing limited reasoning. The answer addresses the question with significant omissions. The candidate has limited use of scientific conventions and vocabulary.

0 marks

The candidate does not make any attempt or give a relevant answer worthy of credit.

A good answer would therefore include:

Advantages:

- Allows production of complex proteins or peptides which cannot be made by other methods.
- Production of medicinal products, e.g. human insulin, factor VIII clotting factor. These are far safer than using hormones extracted from other animals or from donors. Many people with haemophilia within the UK were infected with HIV during the 1980s from contaminated factor VIII extracts.
- Can be used to enhance crop growth – GM crops.
- GM bacteria have been used to treat tooth decay as they outcompete the bacteria which produce lactic acid that leads to dental caries.

Disadvantages:

- It is technically complicated and therefore very expensive on an industrial scale.
- There are difficulties involved in identifying the genes of value in a huge genome.
- Synthesis of required protein may involve several genes each coding for a polypeptide.
- Treatment of human DNA with restriction enzyme produces millions of fragments which are of no use.
- Not all eukaryote genes will express themselves in prokaryote cells.

Hazards:

- Bacteria readily exchange genetic material, e.g. when antibiotic resistance genes are used in *E. coli* these genes could be accidentally transferred to *E. coli* found in the human gut, or other pathogenic bacteria.
- The possibility of transferring oncogenes by using human DNA fragments thus increasing cancer risks.

Q6 a) If plate is spread with penicillin, only bacteria that contain the plasmid can grow (1)

This confirms uptake of plasmid / enables colonies that contain the plasmid to be selected (1)

The second marker gene (Lac Z) is rendered non-functional if DNA is successfully

inserted into it (1)

and is used to confirm insertion of target gene / enables colonies that contain the plasmid with inserted DNA to be selected (1)

ANY 3

b) The plasmid is cut with a restriction enzyme to open the plasmid (1)

The foreign DNA or gene is cut with the same restriction enzyme to ensure complementary sticky ends (1)

DNA is inserted using DNA ligase enzyme (1)

which joins the sugar-phosphate backbones of the two sections of DNA together (1)

Option A: Immunology and disease

Q1 a) Epidemic is the rapid spread of infectious disease to a large number of people within a short period of time (1)

whereas endemic is a disease occurring frequently, at a predictable rate, in a specific location or population

b) Antigen is a molecule that causes the immune system to produce antibodies against it (1)

whereas an antibody is an immunoglobulin produced by the body's immune system in response to an antigen (1)

c) Bactericidal kill bacteria, (1)

whereas bacteriostatic prevent the growth of bacteria within the body by e.g. inhibiting protein synthesis (1)

Examples e.g. tetracycline (bacteriostatic) and penicillin (bactericidal) both for 1 mark

d) Passive immunity occurs when the body receives antibodies, either naturally (e.g. from mother's milk or via the placenta) or artificially from an injection where rapid protection is needed (1)

whereas active immunity occurs when the body produces its own antibodies in response to antigens being present either through exposure to infection or a vaccine (1)

Passive immunity provides immediate protection, but the protection is short-lived because the body has not produced memory cells (1)

Active immunity protects against reinfection where the antigens on the invading microorganism are the same (1)

Q2 a) Prevents synthesis of mRNA (1)

mRNA is not translated at ribosome (1)

proteins are not formed/proteins synthesis is inhibited (1)

b) Both inhibit protein synthesis (1)

Tetracycline is bacteriostatic whereas rifampicin is bactericidal (1)

Tetracycline binds reversibly to the 30S subunit of the bacterial ribosome blocking tRNA attachment whereas rifampicin inhibits {transcription/mRNA production} (1)

c) Use of several antibiotics reduces length of treatment (1)

which reduces risk of antibiotic resistance arising (1)

Q3 a) Primary response (1)

Initial rise in antibody levels following short latency period (1)

b)

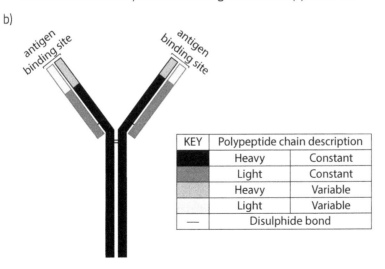

KEY	Polypeptide chain description	
	Heavy	Constant
	Light	Constant
	Heavy	Variable
	Light	Variable
—	Disulphide bond	

Y shaped molecule (1)

Antigen binding site labelled (1)

Heavy/light chains correctly identified (1)

 c) Antigen presenting cells (including macrophages) carry out phagocytosis and incorporate foreign antigen into their cell membranes (1)

  T helper cells detect these antigens and secrete cytokines, which stimulate B cells and macrophages (1)

  B cells are activated and undergo clonal expansion to produce plasma cells and memory cells (1)

  Plasma cells secrete antibodies (1)

  Memory cells remain in the blood to protect against reinfection (1)

  ANY 4

 d) IgM is produced {rapidly after infection/following onset of symptoms} to fight the infection (1)

  After 10 days levels of IgM decrease rapidly to zero after 35 days so is not involved in providing longer-lasting immunity (1)

  IgG is not produced until after 7 days, but is produced at much higher levels / ref to 2× antibody levels (1)

  Levels decrease slightly after 28 days but plateau at 42 days remaining at high levels suggesting IgG provide longer-lasting protection against reinfection (1)

  IgM antibodies are only found in blood and lymph whereas IgG is found in all body fluids suggesting wider protection (1)

Option B: Human musculoskeletal anatomy

Q1 a) Connective tissue

 b) (Yellow elastic cartilage)

  Chondrocytes are surrounded by dense elastic fibres and collagen (1)

  making it elastic but able to retain its shape in structures such as the external ear/pinna (1)

  (fibrocartilage)

  Collagen is arranged into dense fibres increasing the tensile strength (1)

  making it suitable for use in intervertebral discs (1)

 c) The absence of nerves and blood vessels (1)

  means that if damaged, cartilage takes a long time to heal because nutrients have to diffuse into the matrix (1)

Q2 a)

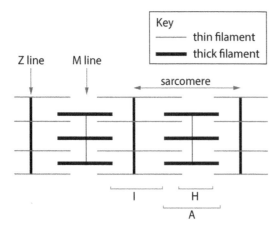

 b) I band and H zone shorten (1)

  A band remains the same length (1)

  Allow consequential error if incorrectly identified in part a)

 c) ATP at the end of the myosin head is hydrolysed into ADP and Pi which are released allowing actin filament to slide (1)

  ATP attaches to the myosin head breaking the cross-bridge unbinding myosin from actin filament (1)

More ATP is hydrolysed to ADP and Pi and a cross-bridge forms with the thin filament further along (1)

d) Calcium channel blockers prevent inward movement of calcium ions (1)

so they are unable to bind to troponin changing its shape (1)

preventing tropomyosin from changing position and exposing the myosin binding sites on the actin (1)

Smooth muscle in walls of arteries relax causing them to dilate (1)

Q3 a) A rigid, movable structure that pivots about a fixed position known as the fulcrum (1)

b) $F_2 = \dfrac{F_1 \times d_1}{d_2}$

$F_2 = \dfrac{196 \times 0.38}{0.05}$ (1)

$= 1{,}490$ N (acc 1,489.6 N) (1)

−1 if no units

c) $F_1 = \dfrac{F_2 \times d_2}{d_2}$

$= \dfrac{2500 \times 0.05}{0.38}$ (1)

$= 329$ N (acc 328.9 N) / 9.8 = 33.6 kg (acc. 33.57) (1)

−1 if no units, allow consequential error

Q4 a) Slow twitch have more mitochondria than fast twitch (1)

Slow twitch are adapted for aerobic respiration whereas fast twitch are adapted for anaerobic respiration (1)

Slow twitch have a high resistance to fatigue whereas fast twitch have lower resistance to fatigue (1)

Slow twitch are adapted for continuous extended contraction whereas fast twitch generate short bursts of strength/speed (1)

ANY 2 – comparison needed

b) Capillary network increases (1) so more blood allows more oxygen, so more aerobic respiration (1)

OR

Increase in number/size of mitochondria (1) so more aerobic respiration (1)

OR

Increase in amount of myoglobin (1) myoglobin is an oxygen store so more aerobic respiration (1)

c) The body relies initially upon creatine phosphate stores (1)

Creatine phosphate releases its phosphate as oxygen levels fall allowing ADP to be phosphorylated allowing for intense bursts of activity (1)

Option C: Neurology and behaviour

Q1 a) i) A = Visual cortex / occipital lobe shaded and labelled (1)

ii) B = Parietal lobe shaded and labelled (1)

iii) C = Frontal lobe shaded and labelled (1)

b) i) Sympathetic nervous system which generally has excitatory effects on the body, e.g. increasing heart and ventilation rates (1) whereas the parasympathetic nervous system generally has an inhibitory effect on the body, e.g. decreasing heart and ventilation rates (1)

Most synapses in the sympathetic nervous system release noradrenaline as the neurotransmitter (1) whereas in the parasympathetic nervous system, acetylcholine is the neurotransmitter (1)

Need comparison

ii) Cerebrum contains two hemispheres responsible for integrating sensory functions and initiating voluntary motor functions / ref to source of intellectual function in humans (1)

whereas the cerebellum is the part of the hindbrain that coordinates the precision and timing in muscular activity, contributing to equilibrium and posture, and to learning motor skills (1)

Need comparison

c) i) Stress event caused a fourfold increase in blood cortisol / ref to increase from 4 to 16 ng/ml (1)

Cortisol is released from the adrenal glands which is controlled by the hippocampus (1)

this resulted in a threefold increase in blood glucose / ref to inc from 110 to 305 mg/dl after one hour (1)

After one hour cortisol levels begin to decrease / ref to 8 ng/ml (1)

which is caused by cortisol binding to glucocorticoid receptors on the hippocampus inhibiting further release (1)

ii) Only one rat was used which decreases reliability of results (1)

Ref to more frequent measurements being taken e.g. every 15 mins (1)

Repeating with more rats / use of control group (1)

Q2 You are not awarded a tick per point, but rather an assessment of your answer is made awarding three main bands. Awarding a mark within the band will depend on how fully you meet the statement.

7–9 marks

Indicative content of this level is...

Detailed description of the different types of behaviour and how they are important to organisms in protecting themselves, finding food, reproduction and developing skills.

The candidate constructs an articulate, integrated account, correctly linking relevant points, such as those in the indicative content, which shows sequential reasoning. The answer fully addresses the question with no irrelevant inclusions or significant omissions. The candidate uses scientific conventions and vocabulary appropriately and accurately.

4–6 marks

Indicative content of this level is...

Description of the different types of behaviour and how they are important to organisms in most of the areas asked, e.g. protecting themselves, finding food, reproduction and developing skills.

The candidate constructs an account correctly linking some relevant points, such as those in the indicative content, showing some reasoning. The answer addresses the question with some omissions. The candidate usually uses scientific conventions and vocabulary appropriately and accurately.

1–3 marks

Indicative content of this level is...

Basic description of the different types of behaviour and how they are important to organisms in some of the areas asked, e.g. protecting themselves, finding food, reproduction and developing skills.

The candidate makes some relevant points, such as those in the indicative content, showing limited reasoning. The answer addresses the question with significant omissions. The candidate has limited use of scientific conventions and vocabulary.

0 marks

The candidate does not make any attempt or give a relevant answer worthy of credit.

A good answer would therefore include:

Behaviour may be either innate (inborn) which is instinctive, or learned.

1. Innate behaviour is more significant in animals with less complex neural systems as they are less able to modify their behaviour as a result of learning. It includes:

● Reflexes are rapid and automatic and protect part of an organism from harm.

● Kineses are more complex than reflexes and involve the movement of the whole organism, and are non-directional, resulting in a faster movement or a change in direction.

● Taxes involve the whole organism moving in response to a stimulus, where the direction of the movement is related to the direction of the stimulus either towards or away from it. An example is seen in woodlice which show negative phototaxis by moving away from light.

2. Learned behaviour builds upon and modifies existing knowledge resulting in a relatively permanent change in behaviour or skills.

- Habituation involves learning to ignore stimuli because they are not followed by either reward or punishment.

- Imprinting occurs at a very early age in a critical period of brain development in birds and some mammals. Konrad Lorenz noted that the young of birds, and some mammals, respond to the first larger moving object they see, smell, touch or hear. They attach to this object and the attachment is reinforced by rewards, e.g. food.

- Associative behaviours include classical and operant conditioning, in which animals associate one type of stimulus with a particular response or action:

  - Classical conditioning involves the association between a natural and an artificial stimulus to bring about the same response. Ivan Pavlov conducted experiments with dogs in which he used a 'neutral stimulus' of a bell ringing, which the dogs learned to associate with food. The dogs would salivate in response to the bell even in the absence of food.

  - Operant conditioning involves the association between a particular behaviour and a reward or punishment. BF Skinner conducted experiments with mice where they learned to press a lever to receive food (reward) or to stop a loud noise (punishment).

  - Latent (exploratory) learning is not directed to satisfying a need or obtaining a reward. Many animals explore new surroundings and learn information which, at a later stage, can mean the difference between life and death.

  - Insight learning does not result from immediate trial and error learning but may be based on information previously learned by other behavioural activities. Kohler conducted experiments with chimpanzees in the 1920s where they were given food but it was out of reach. Chimpanzees were given sticks and boxes, and eventually they learned to use them to reach the food.

Unit 3 Practice paper answers – Energy, Homeostasis and the Environment

Q1 a)

b) Addition of a phosphate group or ion to a molecule (1)

which makes the glucose molecule more reactive and easier to split (1)

by lowering the activation energy (1)

c)

Feature	Mitochondria	Chloroplasts
Mechanism	Uses energy carried by electrons to pump protons across the membrane, they then flow back through stalked particles	Uses electron energy to pump protons across the membrane, which then flow back through stalked particles
Enzyme involved	ATP synthetase	ATP synthetase
Proton gradient	From inter-membrane space to matrix	From thylakoid space to stroma
Site of electron transport chain	Cristae	Thylakoid membrane
Co-enzyme involved	FAD, NAD	NADP
Terminal electron acceptor	Oxygen and H^+	NADP and H^+ (non-cyclic photophosphorylation) and chlorophyll$^+$ (cyclic photophosphorylation)

Unit 3 Answers

Q2

a) Light-dependent – grana or thylakoid membranes clearly labelled (1)

Light-independent – stroma clearly labelled (1)

b) Large surface area (1)

Able to move within palisade cells (1)

c) Non-cyclic involves both photosystems I and II whereas cyclic only involves photosystem II (1)

Non-cyclic produces 2 ATP and NADP whereas cyclic only produces 1 ATP and no NADP (1)

Non-cyclic involves photolysis liberating oxygen whereas cyclic does not (1)

Electrons take a linear path in non-cyclic whereas they take a cyclical path in cyclic (1)

ANY 3

Q3

a) All three correctly labelled 2 marks

Two correctly labelled 1 mark

b) Wider afferent arteriole than efferent arteriole, which creates a higher blood pressure than normal (1)

Capillary epithelium has pores / fenestrae which resist movement of filtrate (1)

Basement membrane acts like a sieve (1)

Wall of Bowman's capsule is made up of {highly specialised epithelial cells / podocytes} through which filtrate passes / ref to pedicels (1)

ANY 3, but structure and function both need to be referenced

c) i) Sodium ions are reabsorbed in the proximal convoluted tubule (1)

by active transport/reference to cotransport with glucose or amino acids (1)

Sodium ions are pumped out of the ascending limb of loop of Henle / ref counter current multiplier into medulla (1)

ii) Water is absorbed from proximal convoluted tubule (1)

in similar proportions to sodium ions or WTTE (1)

iii) Distal convoluted tubule/collecting ducts become more permeable to water (1)

resulting in increased water reabsorbed by osmosis (1)

Ref to aquaporins (1)

which decreases concentration sodium ions in the blood (1)

MAX 3

Q4

a) To establish {baseline / resting urine production} / allow rat to acclimatise (1)

b) Title (1)

Correct axis with units labelled with time on horizontal axis (1)

Correct plots (2) –1 for each incorrect plot

Origin of zero for both axis or linear scales (1)

c) Decrease in urine production seen 10 minutes after injection (1)

Increase in urine production seen after 35 mins/25 mins after ADH injection (1)

Effect of ADH is temporary (1)

Rate of urine production has not returned to original level by 45 min / 40 min after Ref to % decrease or reduction in rate of 3.6 mm^3 min^{-1} (1)

This is due to increased water absorption by distal convoluted tubule and collecting ducts / ref aquaporins (1)

d) Repeat with at least 5 (or more) different rats of similar age/breed (1)

Extend experiment until baseline/4.5 mm^3 min^{-1} reached (1)

Record urine production more frequently, e.g. every minute (1)

ANY 3

Q5 a) During first 4 hours no change in glucose concentration or bacterial cell number (1)

Between 4 and 16 hours the bacterial cell number increases in number as the glucose concentration falls (1)

Between 14 and 16 hours the bacterial cells increase more slowly in number but glucose concentration continues to decrease at the same rate (1)

After 16 hours glucose concentration continues to fall but cell number remains constant (1)

ANY 3

b) Growth ceases due to build-up of toxic waste materials (1)

Stationary phase reached where growth = death (1)

but respiration continues as glucose continues to be used up (1)

c) $\dfrac{5.2 - 2.1 \ (1)}{0.6 \ \ \ \ (1)}$

Allow consequential error / correct reading from graph

= 5 generations, accept 5.2 (1)

d) Decline in number of bacterial cells due to low concentration of glucose (1)

Build-up of toxic waste products (1)

Death phase / death exceeds growth (1)

Q6 a) ETC – {Inner membrane/cristae} of mitochondrion (1)

Krebs – mitochondrial matrix (1)

b) Electrons lose energy resulting in protons\H^+ being pumped from matrix into intermembrane space (1)

Protons/H^+ accumulate / ref to proton gradient (1)

H^+ flow back into matrix through ATPase/stalked particle (1)

Phosphorylating {ADP / ADP + Pi } to ATP (1)

Ref chemiosmosis (1)

ANY 4

Plus

NADH donates protons at first proton pump so more protons pumped across, whilst FADH donates protons at Co Q <u>after</u> first pump so fewer protons pumped across or WTTE (1)

Resulting in 3 ATP per NADH, and only 2 per FADH (1)

MAX 6

c) When succinate is oxidised to fumarate it yields 1 FADH which will generate 2 ATP molecules in electron transport chain (1)

When malate is oxidised to oxaloacetate it yields 1 NADH which will generate 3 ATP molecules in electron transport chain (1)

ATP is only generated in Krebs cycle when α keto glutarate is oxidised to succinate (1)

d) Inhibition results in less fumarate produced and so less malate (1)

so concentration of oxaloacetate decreases (1)

Inhibition stops allowing conversion of succinate into fumarate again (1)

Ref to end product inhibition / negative feedback (1)

Decreased production of {NADH / FADH} (1)

Benefit – prevents build-up of oxaloacetate which may be toxic (1)

Q7 You are not awarded a tick per point, but rather an assessment of your answer is made awarding three main bands. Awarding a mark within the band will depend on how fully you meet the statement.

7–9 marks

Indicative content of this level is...

Detailed explanation of all measures used in the UK to reduce emissions, and how planetary boundaries have been affected as a result.

The candidate constructs an articulate, integrated account, correctly linking relevant points, such as those in the indicative content, which shows sequential reasoning. The answer fully addresses the question with no irrelevant inclusions or significant omissions. The candidate uses scientific conventions and vocabulary appropriately and accurately.

4–6 marks

Indicative content of this level is...

Explanation of most measures used in the UK to reduce emissions, and how planetary boundaries have been affected as a result.

The candidate constructs an account correctly linking some relevant points, such as those in the indicative content, showing some reasoning. The answer addresses the question with some omissions. The candidate usually uses scientific conventions and vocabulary appropriately and accurately.

1–3 marks

Indicative content of this level is...

Basic explanation of some measures used in the UK to reduce emissions, and how planetary boundaries have been affected as a result

The candidate makes some relevant points, such as those in the indicative content, showing limited reasoning. The answer addresses the question with significant omissions. The candidate has limited use of scientific conventions and vocabulary.

0 marks

The candidate does not make any attempt or give a relevant answer worthy of credit.

A good answer would therefore include:

Measures:

- More intensive farming to produce more food more efficiently, use of pesticides to reduce crop losses to disease
- Increased recycling – move to more circular economy, reducing energy needed to produce new goods, e.g. recycling aluminium more efficient than electrolysis of aluminium ore
- Increased energy extraction from waste, reducing need to burn fossil fuels
- Increased use of biofuels – they absorb carbon before being burnt, e.g. biogas, biodiesel, bioethanol
- Ref wind farms / tidal barrages / green fuels / electric cars, etc.
- Managing woodlands more sustainably, e.g. coppicing, selective cutting

Planetary boundaries affected:

- Climate change, biosphere integrity, land system change, biogeochemical flows boundary all still crossed meaning further events unpredictable
- Climate change boundary will have been reduced due to decreased emissions
- Land system change boundary adversely affected by growing biofuels, but positively affected by afforestation schemes/return to more sustainable farming practices
- Biogeochemical flows boundary adversely affected by more intensive farming practices, growing more biofuels.
- Monoculture decreases biodiversity impacting biosphere integrity boundary

Q8   a)   Reference to use of inoculating loop which has been sterilised before inoculation (1)

Loop used to streak bacteria across surface of agar (1)

Reference to sterilising / reflaming loop between each set of streaks (1)

Reference to cooling loop before use (1)

{Keep lid as close to plate / work close to Bunsen burner or flame} during streaking to prevent entry of airborne bacteria/spores (1)

Ref to several lines per streak set / streaks must overlap between sets / no overlap between first and last streak set (1)

ANY 4

b)   i)   Antibiotics I and II are not effective against Gram-positive bacteria as no zone of inhibition is seen (1)

Antibiotic IV is not effective against Gram-negative bacteria as no zone of inhibition is seen (1)

Antibiotic III is effective against both bacteria, but antibiotic III is more effective against Gram-negative because the zone of inhibition is larger (accept converse) (1)

ii)   Antibiotic IV (1)

Penicillin is only effective against Gram-positive bacteria (1)

Cannot be antibiotic III as this is also effective against Gram-negative bacteria (1)

Penicillin inhibits formation of cross-linkages in peptidoglycan cell wall (1)

Unit 4 Practice paper answers – Variation, Inheritance and Options

Q1   a)   i)   A = Seminal vesicle (1)

Secrete mucus/excretion to aid sperm mobility

ii)   B = Prostate (not prostrate) gland (1)

Produces alkaline secretion to neutralise acidity of urine (1)

iii)   C = Vas deferens (1)

Carries spermatozoa from testis to urethra/prostate gland (1)

b)   Ref to increased temperature of testis qualified, e.g. wearing tighter underpants, hot baths / electromagnetic radiation from computer equipment / stress / alcohol / drug use or smoking (1)

Reduced {fertility / birth rates} /decreasing population size / increased uptake of IVF (1)

c)   i)   Cell X = spermatids

ii)   {Secondary spermatocytes/ Cell X}, spermatids, spermatozoa (all needed for 1 mark)

iii)   Division between spermatogonia and primary spermatocytes involves mitosis whereas division between primary and secondary spermatocytes involved meiosis I (1)

Division between spermatogonia and primary spermatocytes maintains chromosome number whereas division between primary and secondary spermatocytes halves chromosome number (1)

Division between spermatogonia and primary spermatocytes produces diploid cells whereas division between primary and secondary spermatocytes produced haploid cells (1)

iv)   Must link structure for function, i.e. nucleus contains haploid number of chromosomes which restores diploid cell at fertilisation (1)

Mitochondria in mid piece produce ATP for locomotion / tail for locomotion (1)

Acrosome contains proteases to digest cells of corona radiata / zona pellucida (1)

Q2

a) Correctly labelled amino acid showing

NH_2 group labelled as amino group (1)

attached to central carbon atom with R group (1)

attached to COOH group labelled as carboxyl group (1)

b) i) Change in the base sequence of DNA / one or more DNA bases in a gene changes (1)

by {addition/substitution/deletion} (1)

resulting in a change in the order of amino acids in the protein (1)

leading to a change in the tertiary structure of the protein (1)

ii) Change may occur in non-coding region/intron (1)

Change may be silent / may not change order of amino acids / ref to several codons for each amino acid (1)

Amino acid change may not alter structure/function of protein (1)

c) i) {PKU gene must be on the autosomes / not sex linked} as males and females are both affected (1)

Must be recessive, e.g. parents 1 and 2 do not have PKU but child 5 does, so both must be carriers (1)

Answer must be qualified to gain mark

ii) Chance of male child is 50%/50:50 (1)

Chance of a child having PKU is 25%/1:4 (1)

as it is a recessive condition (1) accept consequential error if part i) incorrect

ANY 2

Chance of male PKU child is therefore 12.5% or 1:8 (1)

MAX 3

d) Design primers to section of mutated DNA (1)

Add to DNA polymerase, buffer and deoxyribonucleotides (1)

Run at three different temperatures (accept any in range of) 50–60 °C, and 70 °C and 95 °C (1)

Repeat for 30–40 cycles (1)

Q3

a) Both share same genus *Chaetodon* (1)

but are different species, so cannot interbreed to produce fertile offspring (1)

show similar morphology/characteristics (1)

may show different mating behaviour (1)

b) Genetic fingerprinting /microsatellite analysis (1)

DNA hybridisation using DNA probes, and visualising by gel electrophoresis (1)

c) Ref to {allopatric speciation / geographical isolation} when isthmus formed, separating two populations (1)

this prevented interbreeding (1)

Ref different environmental conditions in Pacific and Atlantic oceans (1)

which led to different selection pressures (1)

Ref which led to changes in morphology/breeding (1)

and two distinct gene pools which could not mix (1)

ANY 5

Q4

a) Use of appropriate letters with capital for dominant feature, and correct parental genotypes e.g. BbVv vs bbvv (where B is brown body, b is black body, V is normal wing, v is vestigial. (1)

Correct gametes (1)

F1 genotypes correct (1)

Expected phenotypic ratio is 1 brown-bodied fly with long wings to 1 brown-bodied fly with vestigial wings to 1 black-bodied fly with long wings to 1 black-bodied fly with vestigial wings (1 mark for phenotypes, 1 for ratio)

b)

Category	Observed (O)	Expected (E)	O – E	$(O–E)^2$	$(O–E)^2/E$
Brown body, long wings	26	**15**	**11**	**121**	8.07
Brown body, vestigial wings	6	15	–9	81	5.40
Black body, long wings	5	15	–10	100	6.67
Black body, vestigial wings	23	15	8	64	4.27
Σ	60	60			24.41

$\chi^2 = 24.21$ (1)

c) The null hypothesis is that there is no significant difference between the observed and expected values. (1)

Because the calculated value 24.41 is <u>more</u> than the critical value at p = 0.05, 7.82, we can reject the null hypothesis, so any differences between observed and expected results seen were not due to chance (1)

Body colour and wing length genes are linked, which means they are located very close to each other on the same chromosome (1)

The consequence of this is that these alleles are much less likely to segregate independently to form gametes (1)

Q5

a)

Concentration of PBZ / mg dm⁻³	Number of cress seedlings germinated	Percentage germination / %	Height of seedlings / cm	Average height of seedlings / cm
0	10	**83.3**	8.1, 7.9, 8.0, 7.5, 8.4, 7.5, 8.2, 6.1, 5.9, 8.9	**7.7** (acc 7.6)
10	10	**83.3**	5.6, 5.5, 6.4, 6.0, 8.0, 5.3, 5.2, 5.7, 4.9, 5.0	**5.8** (acc 5.7)
50	5	**41.7**	3.3, 6.9, 3.4, 2.9, 3.1, 0.0, 0.0, 0.0, 0.0, 0.0	**2.0**
90	0	**0.0**	0.0, 0.0, 0.0, 0.0, 0.0, 0.0, 0.0, 0.0, 0.0, 0.0	**0.0**
		(1)		(1)

1 mark for each column correct

b) IV = Concentration of PBZ / mg dm⁻³

DV = Number of cress seedlings germinated <u>and</u> height of cress seedlings / cm

c) You are not awarded a tick per point, but rather an assessment of your answer is made awarding three main bands. Awarding a mark within the band will depend on how fully you meet the statement.

7–9 marks

Indicative content of this level is...

Detailed conclusions from the experiment, commenting fully on the accuracy of the results, and suggesting improvements.

The candidate constructs an articulate, integrated account, correctly linking relevant points, such as those in the indicative content, which shows sequential reasoning. The answer fully addresses the question with no irrelevant inclusions or significant omissions. The candidate uses scientific conventions and vocabulary appropriately and accurately.

4–6 marks

Indicative content of this level is...

Drawing most conclusions from the experiment, commenting on the accuracy of the results, and suggesting some improvements.

The candidate constructs an account correctly linking some relevant points, such as those in the indicative content, showing some reasoning. The answer addresses the question with some omissions. The candidate usually uses scientific conventions and vocabulary appropriately and accurately.

1–3 marks

Indicative content of this level is...

Basic conclusions from the experiment, commenting partially on the accuracy of the results, and suggesting one improvement.

The candidate makes some relevant points, such as those in the indicative content, showing limited reasoning. The answer addresses the question with significant omissions. The candidate has limited use of scientific conventions and vocabulary.

0 marks

The candidate does not make any attempt or give a relevant answer worthy of credit.

A good answer would therefore include:

PBZ inhibits germination of cress seedlings at concentrations <u>above</u> 10 mg dm^{-3}

PBZ reduces height of seedlings at a concentration of 10 mg dm^{-3} ref to amount, e.g. by average of 1.9 cm / 25%

50 mg dm^{-3} of PBZ caused a reduction in germination by 50%

50 mg dm^{-3} of PBZ caused a reduction in seedling height by average 4.2 cm /5 5%

PBZ prevents synthesis of GA12 aldehyde a precursor needed to make Gibberellic acid (1)

Without GA, genes involved in transcription and translation of amylases and proteases aren't switched on, so food stores aren't mobilised

Reliability:

Ref to anomalous results which lie outside mean, e.g. at 50 mg dm^{-3} 6.9 cm, or at 10 mg dm^{-3} 8.0 cm

These results affect mean result suggesting result may not be reliable

Improvements:

Ensure seeds are same age

Measure volume of PBZ administered

Not just repeat with more seeds unqualified

Q6 Option A

a) Humoral (1)

b) Four (1)

Two light and two heavy / or reference to alpha and beta (1)

c) 15 – 5 = 10

10/5 × 100 (1)

= 200% (1)

−1 no units

d) Engulf foreign {cells/virus/bacteria} into a vesicle/phagosome (1)

Secrete lysozymes to digest contents (1)

e) i) Loss of helper T cells / macrophages (1)

T helper cells stimulate phagocytosis, antibody production and activate T killer cells (1)

Phagocytes do not engulf {virus/bacterial/named infection} (1)

Antibodies are not produced (1)

Killer T cells do not bind to foreign cells and so they are not destroyed or WTTE (1)

MAX 4

ii) Both primary and secondary responses provide low concentrations of antibodies in the blood (1)

Secondary response does not produce a significant difference from the first / ref to antibody concentration in the blood being similar (1)

due to T helper cells not stimulating antibody production (1)

Ref to vaccination not conferring protection but might cause side effects (1)

f) i) Prevent addition of nucleotides in reverse transcription of viral genome (1)

Viral DNA not produced (1)

Viral DNA is not incorporated into cell's DNA (1)

ii) HIV does not have a cell wall containing peptidoglycan (1)

Penicillin prevents formation of peptidoglycan cross links/formation of peptidoglycan cell wall (1)

Q7 Option B

a) Connective (1)

b) Hyaline (1)

c) Hinge (1)

d) White elastic cartilage / fibrocartilage (1)

Collagen is arranged into dense fibres (1)

which increases tensile strength for support/protection of spinal cord (1)

e) $F_2 = \dfrac{(70 \times 9.8) \times 40}{3}$ (1)

$= 27{,}440 / 3 = 9{,}147\,N$ (1)

-1 no units

f) i) Respire anaerobically for short periods so do not need good oxygenation (1)

Creatine phosphate releases its phosphate as oxygen levels fall allowing ADP to be phosphorylated (1)

allowing for intense bursts of activity anaerobically (1)

High levels of myosin ATPase enable more cross bridges to form between actin and myosin {per unit time/in the same time} (1)

ii) Respire aerobically for extended periods so don't rely on anaerobic respiration (1)

therefore, don't need stores of creatine phosphate to release phosphate (1)

Muscles contract more slowly with less force so require fewer fibres (1)

g)

Type of training	Effect of training	Advantage
Endurance training	Increase in number and size of mitochondria	More aerobic respiration possible
Endurance training	Capillary network increases	Increased blood supply to muscle results in more oxygen so more aerobic respiration
Weight training	Increase in number of myofibrils and size of muscles	Increases strength
Endurance training	Increase in amount of myoglobin	Myoglobin is an oxygen store so more aerobic respiration
Weight training	Increase in tolerance to lactic acid	More anaerobic respiration possible

Must link effect of training to benefits (1 mark for each correct)

Q8 Option C

a) Hypothalamus correctly labelled (1)

b) Homeostasis (1)

Ref to osmoregulation / releases ADH through pituitary gland (1)

Ref to maintaining body temperature (1)

Controls pituitary gland (1)

Regulates behaviours such as sleep (1)

Ref to regulating thirst (1)

Controls autonomic nervous system (1)

ANY 3

c) Correct area shaded e.g.

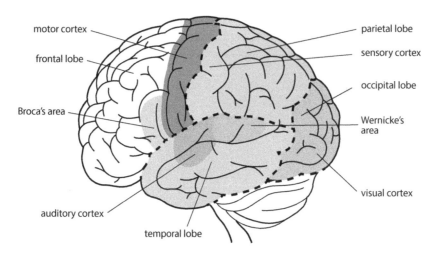

d) Innervates (supplies nerves to) the muscles (1)

e.g. intercostal muscles and muscles in mouth, and larynx required to produce sound (1)

e) i) Reduces number of synapses per neurone / ref to decreasing from 15,000 per neurone to 1000 –10,000 (1)

Creates hard-wired connections allowing quicker and more accurate transmission of signals (1)

ii) Inhibition of GABA receptors triggers {synaptic pruning /reduction of number of synapses per neurone} (1)

so could be a treatment for schizophrenia caused by abnormal spine density (1)

Caution – experiments were in mice, not humans, so the effect may not be the same (1)

f) i) Left for 5 mins – to allow woodlice time to acclimatise to new environment (1)

Rotated – to reduce effect from any stray light (1)

ii) Negative photoaxis (1)

Movement is related to direction of stimulus / move away from light (1)

iii) Experimental error (1)

Woodlice searching for food (1)

Effect of stray light / rotating chamber (1)

ANY 2

iv) {Control/ monitor} light intensity with example, e.g. monitoring with light meter / blackout lab and shine single light source on dish (1)

{Control/monitor} temperature with example e.g. use heat shield for light source / monitor with thermometer (1)

Use woodlice of same age/size (1)

ANY 2